AUTO-OFF @ 65

WHAT IF HUMANS ARE BORN WITH A FIXED EXPIRY DATE?

NEERAJ GUTGUTIA

BlueRose ONE.com
Stories Matter
NewDelhi • London

BLUEROSE PUBLISHERS
India | U.K.

BlueRose ONE.com
Stories Matter
NewDelhi • London

For permissions requests or inquiries regarding this publication, please contact:

BLUEROSE PUBLISHERS
www.BlueRoseONE.com
info@bluerosepublishers.com
+91 8882 898 898
+4407342408967

ISBN: 978-93-6783-546-3

Cover design: Daksh
Typesetting: Tanya Raj Upadhyay

First Edition: January 2025

About the Author

Neeraj Gutgutia is a 45-year-old technology entrepreneur and founder of two companies. Right2Vote.in is a Government of India approved online election platform created to ensure the 'right to vote' for all. Time2Justice.ai is an AI-powered legal platform designed to speed up the Indian judicial system. Neeraj hopes to experience a developed India in his lifetime. His two ventures and this book are his small contribution towards this mission.

Neeraj is a qualified Chartered Accountant (CA) and Master of Business Administration (MBA) from IIM Indore. He also studied at ISB Hyderabad, Stanford University, California and Cornell University, New York. He has worked with companies like Reliance Industries Limited, ICICI Bank, and Hiranandani Group. He also briefly worked with Businessworld magazine and was the founder and editor of the 'I Magazine' at IIM Indore.

Neeraj is interested in Technology, Politics, Psychology, Economics, and Finance. He loves to drink cocktails made of all of the above. This book is a special cocktail he has created for you, and he hopes you will enjoy it.

Neeraj lives in Mumbai with his wife and two lovely daughters.

LINKEDIN: https://www.linkedin.com/in/neeraj-gutgutia/
WEBSITES: www.gutgutia.in, https://right2vote.in/, https://time2justice.ai/
EMAIL: neeraj@gutgutia.in

Disclaimer

This book is a work of science fiction. While some characters may be loosely inspired by real-life individuals, they are purely fictionalised for the narrative. The author has crafted these characters with unique attributes and story arcs to serve the plot's requirements. It should be noted that no character, organisation, or institution depicted in this book represents real-world entities, as the entirety of the content is a product of fiction.

Table of Contents

Chapter 1:
The Big Announcement

The President

"You all will die at 65!" Announced the new President of the USA.

Today is 20th January 2025. Micky Cash, the 51st President of the United States of America is being sworn into office. The President is a very colourful and controversial character. He won the election which means he has a great fan following in certain sections of the American Population. However, he has an equally large group of haters who cannot stand him.

He was also the 49th President of the USA. He narrowly lost the bid to become the 50th President of the USA post in the 2020 elections. To an extent, COVID-19 and his handling of COVID-19 as President of the USA was one of the major reasons for his loss. But he and his followers did not take that well. His fans and followers stormed the Capitol in January 2021 during the swearing-in ceremony of his opponent. United States Secret Service Personnel had to be called in to control the situation. The situation closely resembled a coup, a situation which is generally associated with autocratic countries or failed democracies. For America, which prides itself on being the oldest democracy in the world, it was a very shameful moment. Later, Micky was pronounced guilty of conspiring to overturn the democratically elected government and of interfering with the 2020 election results. However, he has been given a delayed sentence by the courts and the first thing Micky is expected to do after the swearing-in ceremony today is to give himself a Presidential pardon.

Conspiring to overturn the election results is not the only indictment against the President. There are charges relating to government official secrets. There are charges relating to the use of hush money to cover up extramarital affairs. He is well known for his flamboyant lifestyle. During the first election campaign, his voice recording was leaked, where he was caught using disrespectful words about women. He dismissed those talks as petty locker room talks.

He is also a very successful businessman, a billionaire who made his fortune across businesses but primarily in real estate. He has also been a television host, where he plays himself, and true to his character fires apprenticeship candidates at the drop of a hat without any remorse.

The President has an image of being an unpredictable madman who can take immoral, unconventional, and controversial steps to achieve his goals. He is not known to hold himself back, on moral or other grounds, from achieving his goals. At times, his goals can be equally unconventional, controversial, and immoral.

As promised by him, after his election people are expecting some big announcements today. He has promised many things during his election campaign. Today his speech should indicate what he wants to focus on first.

The President started off the speech predictably with a lot of chest-thumping, sloganeering, and congratulating himself and his team. On his side, all his key appointees and aides are standing and cheering whatever he is saying. Among the long list of aides standing behind him are two business czars – Leo Smith and Cyrus Clark.

Leo Smith, aged 53, is considered the biggest genius of our times and owns several businesses across sectors, all built from scratch by him with his unapparelled tenacity, innovation, ambitious vision, and equally mindboggling execution. Many of these businesses are worth hundreds of billions of dollars making him the richest person on Earth. His businesses range from giga solar plants, self-driving electric cars, space travel, satellite communication, and social media platforms to flame throwers, perfume, and even shorts. All his businesses are more ambitious than others and still, he manages to come up with crazier ideas now and then. Whatever new business idea he comes up with, sounds crazy at the start but he has proved himself every time with his execution. From driverless cars to reusable rockets, he has proved the world wrong many times. Even his biggest competitors and rivals like NASA and General Motors, address him as a Genius. If today all 8 billion people across the world were asked to vote for the biggest genius in the world, Leo would surely score more than 80% of the votes. However, the same 80% might also vote him as crazy!

Leo is crazy in many ways. He believes mankind needs to colonize another planet in the solar system so that in the future if there is a sudden extinction event on Earth, as happened with Dinosaurs, mankind can survive by migrating to the other planet. Something like Noah's Ark. He is building his own Noah's Ark to save mankind. He is working towards colonizing Mars and is committed to achieving it by 2028. Leo also believes that to save mankind everybody should have more children. He has eleven offspring from 3 different partners.

He thinks himself to be the ultimate protector of mankind. He thinks himself to be a gift to mankind. In fact, not just a gift to mankind, he thinks of himself to be God. There is an often-spoken joke about him – '*the only difference between God and Leo is that God does not think he is Leo.*' It is only half a joke, with all the impossible Leo has been able to achieve in one lifetime, many people actually compare him to God.

Leo is flamboyant, impulsive, and quick to offend. He is infamous for his impulsive comments on his own social media platform, rebranded by him to 'Y', which has landed him in trouble several times. He has a rockstar-like fan following on social media platforms. Once he bought a social media platform, at a price twice its worth, in a snap, just because he got offended by something.

Cyrus Clark, aged 69, is no less a genius but much older now and much more mild-mannered. Cyrus can be considered the world's first software genius. He is the pioneer of personal computer software interface and to date, more than 90% of the world's computers are still running on his software. It is difficult to imagine computers as we know them without their software. Cyrus is not only the founder of MacroSoftware, one of the 5 biggest tech companies in the world, but also one of the largest shareholders in another big 5 tech companies where he invested as a white knight a few years back when that company was in trouble. He has held the position of the richest person on Earth for the longest period since Fortune started the rankings. He still is one of the top five richest people in the world and has lost his top ranking majorly due to his philanthropic activities. Now, he holds the top ranking in the list of biggest philanthropists in the world. His foundation is working on social causes like the eradication of poverty from the world, education, sanitation, medication for the poorest of the poor, and much more. He is one of the leaders in the fight against pandemics and led the COVID-19 vaccination research from the front.

Cyrus was geeky when he was young. He was known to be a workaholic and hard taskmaster. Although he is never known to be colourful like Micky or flamboyant like Leo, he too is believed to have a weakness for women. There is a rumour that he was forced to give up the operational role in his company MacroSoftware due to his improper advances to a female colleague. Recently he got divorced from his wife of 27 years. Again, unconfirmed rumours suggest it was due to his association with Jeffery Epstein, one of the biggest sex offenders to be born in the USA. It is rumoured that he visited Epstein's infamous island where Epstein used to host celebrities, politicians, and billionaires and used to offer them all kinds of entertainment.

His image has evolved over the years. He now has an image of a billionaire philanthropist working to save mankind from poverty, diseases, and pandemics.

It's been 30 minutes and the content of the President's speech is not very different from any speech of a newly elected President. This was not expected. Considering his image of being an unconventional and unpredictable person, everybody was expecting much more. The crowd is loving the speech as he is egging them in his signature style, the noise level in the stadium is deafening. However, something is missing. We were expecting more. We were expecting some big announcement, some 'world-shattering' announcement to quote his own words. It seems he is done with the speech. People are hooting and shouting like crazy. "USA! USA! USA!" The aides behind the President have started stepping down from the podium. The President is still egging the crowd and enjoying the limelight, again in his signature style. "Micky! Micky! Micky!"

Leo and Cyrus have not stepped down from the podium. They have moved even closer to the President. Even the Vice President and the First lady have stepped down from the Podium. What is happening? Are Leo and Cyrus planning to make a speech after the President's speech? That would be a first! The President is in no mood to step down. Generally, a president would step down into the crowd and shake hands with a few of his supporters, a symbolic gesture to show that he treats himself as an equal and partner of the citizens and not a preacher on a podium. But that does not seem to be happening today. It seems he wants to speak again!

"Friends, the speech I just gave just now was in the capacity of the leader of the most powerful nation of the world. What I am about to speak next is in the capacity of the leader of the world!

Mankind is at risk of extinction!

We are not doing anything about it. On the contrary, we are amplifying it but putting our best minds and big dollars to accelerate the process.

The average global life expectancy is around 70 years and for the USA it's 77 years. Average life expectancy has been increasing consistently over the last few decades and with advancements in modern medicine, especially with the help of Artificial Intelligence (AI), it might soon touch 100 years.

On the other hand, the world's fertility rate has fallen to 2.26, and in the USA, it has fallen to 1.66. In countries like South Korea, it is as low as 1.12. The faster the country's pace of development, the faster the fall in fertility rate. The world needs a minimum fertility rate of 2.2 to maintain the population; however, it is estimated that it will soon fall to less than 1.5.

The average age of the world population is 31 years and for the USA population is 39 years.

It is a well-known fact that the more developed a country is, the higher the average life expectancy, the higher the average age of the population, and the lower the fertility rate. This means, as mankind develops at a faster pace the number will become even more stark!

Average life expectancy increase, after a certain level, is not good news. It is bad news! Very bad news. It is having a direct impact on the fertility rate, there is a perfect inverse relationship! If anybody argues otherwise, he is a dimwit. This is leading mankind to extinction."

The crowd is confused, what is the President talking about, where is he going? The crowd is in the mood for celebration and not in the mood for a lecture on Biology or whatever it is. Energy and noise levels have come down in the stadium but still, it is quite high.

The President finally drops the bomb! This bomb is even bigger than the Atom bomb dropped by President Truman!

*"To save mankind from extinction, we all need to die early. We have a plan for that....**we all will die on our 65th birthday.***

A death, which will save all of us from the misery of old age. A death, after a well-lived full life."

There is a pin-drop silence in the stadium, there is pin-drop silence at Times Square, people across the world are watching this telecast live and there is pin-drop silence in almost every drawing room. People are confused, people are shocked, people can't believe what they just heard.

The President was expecting it. After almost a one-minute pause, which felt like an eternity, he repeats

*"Yes, you heard it right....**you all will die at 65!"***

"Dying at 65 is not a bad thing, actually it is a very good thing.

Is life after 65 worth living? Most of us are bedridden, dependent on daily doses of medicine, suffering from one pain or another.

A life where we have nothing to do. From morning to evening spending time itself is a task.

A life where we are a liability to our kids, sitting in the corner of their homes, dependent on them not only for food but also for help to reach the toilet.

A life where we are a liability to society, sucking up the resources which can be put to much better use by the younger generation.

Let's live life to the fullest till we are 65 and then die happy, without the curse of old age.

And it's not a big deal. The average life expectancy in the world is 70 years. You are not dying 5 years early, but you are avoiding the worst five years of your life.

We are making death predictable so that you can plan your life and live your life to the fullest."

The public is still confused, bewildered, wondering if is it real or some bad dream. Few people have started getting angry but are still not sure how to react or take out their anger.

"I alone cannot take credit for this mankind-saving world-shattering plan. We call this plan – Auto-off @ 65. I have been supported in this endeavour by various world leaders from across the world. Many of them would not like to be named and hence, I am not taking names. However, the two biggest contributors to this plan without whom we would not have been able to achieve this scientific feat are Leo Smith and Cyrus Clark.

We have been working on this goal to save the world since 2020. Since Covid. These were the people who came up with the idea and managed to make it scientifically possible. Without the Satellite Tech of Leo Smith and the Medical Tech of Cyrus Clark, this would not have been possible. Guys, please give a big round of applause for Leo Smith and Cyrus Clark! The biggest geniuses of our generation and all time."

The Satellite Czar

People are like – "Applause! Really?" This was the last nail in the coffin, now people are angry, very angry. The President has left the dais through a secret exit at the back along with United States Secret Service (USSS) personnel. Without people realizing it, the USSS personnel moved in and took full control of the stadium. They were prepared for this! They have been briefed that they would be required. USSS guys are now spread across the stadium. Scores of snipers are visible at vantage positions. Dozens of these USSS guys are on the dais, with their advanced weaponry, facing the public, most people realise it would be foolish to react.

Leo Smith steps forward towards the podium. He is as confident and sure-mannered as ever. As if he is oblivious to the anger brimming up in the crowd. As if, he is about to announce yet another successful launch of a rocket. His demeanour is similar to his demeanour a few days ago when he announced the successful launch and return of his rocket. He is behaving as if he is about to take credit for one of the biggest achievements of his lifetime.

"Guys, I am 53. I have 12 more years to live. I will live it to the fullest. I will live a good fulfilling life without the curse of old age. And we are ensuring the same for all of you. You all will also live a long, fulfilling life, without the curse of old age.

As the President said, Auto-off at 65 is a good thing. This will save you from the curse of old age. This will make death predictable and help you plan for it. But the most important thing, this would let the earth's resources go back to future generations and save mankind."

People in the crowd are now mad with anger. If not for the USSS personnel they would have attacked Leo and Cyrus by now. Leo had a fan following bigger than any sports star or film star. But now the same fans are angry. Angry and scared of Leo, Cyrus, Micky, and the whole gang who planned Auto-off @ 65. Suddenly Leo's image has turned from that of a demi-god to an eccentric and mad businessman willing to massacre people without caring a damn.

People are thinking - who are these billionaires and politicians to decide how long I or my family lives? How can these people kill us? This is a planned massacre. They are calling it saving mankind, they are in reality massacring mankind. Leo continues to speak.

"Please understand, with modern medicine, we have been extending our lives. But when we are extending our lives, we are also reproducing less and reproducing later. We are making the Earth a habitat of the old. Our genes are deteriorating and we are passing on the deteriorating genes to the next generation. Over a few generations, our genes would deteriorate so much that Earth would become the habitat of the diseased. Already our kids are being born with breathing problems, already our kids are needing spectacles at a very early age. We are living with the pandemic of old age which we are not realizing now. We are killing mankind by living long and passing on diseases to the next generation. We need to plan our lives; we need to reproduce at the right age and die at the right age.

Cyrus Clark is an expert on health and medicine and I will leave it to him to explain why Project Auto-off @ 65 was necessary to save mankind. I would spend the next few minutes explaining how we have been able to achieve it and how it would be a painless switch-off and not death.

You are aware that my organization 'SatelliteLink' has launched more than 10,000 satellites into space. These satellites are the reason why all of us across the world can communicate seamlessly from any corner of the world. These satellites have been my gift to the mankind in more than one way. We are using the same satellites also for the execution of project Auto-off @65. The satellites are sending rays on earth, each and every corner of the earth, touching each and every human being. The rays have been used to do the gene editing to ensure Auto-off @65. Auto-off @65 was not possible only with the Satellite rays, an enabler was required in the human body. The enabler was inserted in our bodies via COVID-19 vaccination in 2020, thanks to the ingenuity of Cyrus. I will leave it to Cyrus to delve into it in detail during his speech.

The satellite has completed the job of gene editing and 100% of mankind has been covered. There is no exception. People living in caves or bunkers are also not spared from gene editing. Be rest assured, that we have ensured everybody is treated equally without any exception.

In the long term, the satellites would continue monitoring the mankind from top, taking high-resolution images of everybody to advance and notify us humans if there is any deterioration in human genes. We have been doing this gene imaging for years now, and based on the satellite imaging we realized the need for project Auto-off @ 65.

Please note: the Auto-off @ 65 would be painless like a switch-off of a button. We fear death due to fear of pain. With Auto-off, we have taken away the pain from the process of dying.

And again, if you are still scared of dying @ 65, we have an escape route from Auto-off @ 65! My organization is offering free one-way travel to Mars to people who are willing. The Auto-off @ 65 technology is built in a manner that it is effective only in Earth's environment. On Mars, Auto-off would not apply and you can continue to live till your natural death. Auto-off @ 65 is designed to save the Earth. If you are willing to live on Mars, you are welcome. Anyways, to save mankind from unknown risks, we need to build an alternative habitat. An alternative to our mother Earth, in case we are not able to save her!"

The Modern Medicine Czar

Leo is done with his speech. He has stepped back and left the podium open for Cyrus. People are agitated, confused, bewildered, disillusioned – all at the same time. Different age groups are reacting differently. Younger people still in their twenties or younger are agitated about how politicians can take such calls but they are not really scared. They still have many decades to go before they reach 65. They have never given a thought to how long they will live. They are not sure whether they want to live that long nor they can picture themselves as 65 years old. For them, it is sounding like climate change doomsday predictions, something that will happen many, many years later.

On the other hand, people who are in their early sixties are really scared. They are so scared that it seems like this announcement itself is enough to kill them. It seems they are about to get a heart attack and collapse. Different people are in different stages of grief. Some are still in denial; they are not ready to believe it and still have a feeling that it is just a bad dream. There is yet another group that has passed the stages of denial and anger. They are now in the stage of 'bargain'. They want to ask questions; they want to buy some more time. They want to know what's the start date of project Auto-off @ 65 and if can it be postponed for a few years. There is still another group watching the event on TV sets from their hospital beds or bed-ridden at home. They have already lost their lust for life and want to painlessly escape to death. For them, this announcement is relieving, a bit scary, but at the same time relieving.

There is another group, in their thirties and forties, who have a mixed feeling. They are not sure of their own feelings. They do not want to die, at least do not want to be forced to die but at the same time wondering whether they really want to live beyond 65. Old age and associated medical problems scare them, they want to stay healthy all their life. In a way, this project Auto-off @ 65 is granting their wish. This group is more worried about their parents who have crossed 60. They do not want to lose their parents. The fear of losing their parents is making them angry.

Cyrus steps up to the podium. Unlike Leo, he is not flamboyant. He is soft-spoken and has much more empathy. He is not vain like Leo, who is full of himself and thinks he is a gift to mankind. Cyrus also used to be like that but with age, he has matured and become much more down to earth. Cyrus can sense the mood of the people. The mood has changed from celebration to surprise to anger and now to helplessness. Cyrus is feeling sorry for the people. He understands it is a big shock for them. It is not easy to embrace death for yourself and your loved ones. But he also believes in the higher purpose of saving mankind and has joined the project with a full understanding of the implications. As a matter of fact, the project is the brainchild of Cyrus. Micky and Leo joined the project in 2020. Cyrus has been working on it since 2015!

Cyrus senses that he has to pacify the crowd otherwise it will get out of hand. A riot has not broken out yet in the stadium just because of the presence of USSS personnel. The environment is tense and things can go very bad anytime. Not only in the stadium but across the world. Cyrus commences his speech.

"Dear fellow citizens of the USA and the world, trust us, we are doing this for the good of mankind. Trust us, auto-off at 65 is a good thing. I am 69 years old and I am in the same boat as all of you. There are no exceptions. The same rule applies to all. Let me first tell you how the project will work along with the timelines and then I will get into the technical aspects including gene editing, side effects etc. I know you must be having hundreds of questions in your mind. Let me speak, I will try to answer most of the questions that are coming to your mind during the next 30 minutes. Post that, if you people need any further clarification, feel free to ask us. We will not leave till all your queries and concerns are answered."

Cyrus starts off with the details of the project. It seems he plans to mechanically lay bare all the finer details of the project. It seems he wants to be transparent. With his opening words, he has managed to build some trust. At least people are willing to hear him out.

"The project is already set into motion and there is no going back. We have not created any off button in Project Auto-Off @ 65. The project was set into motion in 2020 during the Covid period. I have been working on the idea and technology to edit genes since 2015. During the Covid pandemic, I was able to convince the President, Leo, and some of the other prominent world leaders about the need for such a project.

Gene editing ensures all humans with automatically switch off as soon as they reach the age of 65. Gene editing also ensures that any human born to a parent with gene editing will also have the modified genes and would auto switch off at 65. I prefer to use the term switch off instead of die as we have managed to ensure a 100% painless process of switching off. The whole process takes less than a second and hence there is no suffering.

For people who are already 65 or more than 60 years old as of today, we have programmed the project in such a manner that they will get another 5 years irrespective of their age. This means till 1 January 2030, there would not be a single case of switch-off. The process of switching off will begin on 1st January 2023. This we have done to ensure everybody gets time to plan their affairs, say goodbye to their family and friends and also complete whatever is pending on their bucket list.

I am 69 years old, hence I will also switch off on 1st January 2030. So will our President Micky. Please note, that there is no exception. The President and I, both of us, would be the first ones to get switched off under the project. You can trust us when I say, that the process would be totally painless and devoid of any suffering."

The crowd is shocked again. Everybody is looking at each other. They are unable to trust their ears again. Did he just say that he and the President will die on the first day of the project? Really?! Some people are happy, some are relieved that at least the project is not discriminative.

People older than 60 took a sigh of relief. At least they have 5 years. After hearing the President's speech, they were worried that they could die at any moment. They have time to plan things, they have some time to spend with their families. Some of the senior citizens who are suffering are thinking: '*Anyway, I won't last another 5 years.*' Or '*anyway, I don't want to last another 5 years.*'

Some younger people in the crowd are now even more confused. How can two of the three main plotters of the blasphemous plan die on day 1? What if something went wrong? Are these people crazy, have they already gone senile? Meanwhile, Cyrus continues to speak, now he wants to justify the motive, after all, it was his idea.

"I am a big advocate and supporter of modern medicine. As you all must be aware, I through my foundation have been funding and closely working on several research aiming to invent lifesaving drugs. I am very proud that my foundation has been the biggest contributor to medical research in the last 20 years.

We have also achieved a lot with these research projects, we managed to eradicate polio from the world, we managed to control Ebola, we have limited HIV, and we have made groundbreaking progress with cancer. We were also the ones who saved the world from the worst pandemic in recent times – Covid. We were the ones who managed to get the vaccine invented in time. So please trust us.

Way back in 2015, we realized that we were over-achieving in our goals and it is creating another problem which is defeating the very goal. It is a big IRONY that by inventing lifesaving modern medicine we pushed mankind towards extinction. It was a great run till we managed to get the average life expectancy up from around 31 years in 1975 to 66 years in 2000. However, by 2015 it crossed 70 years. And with rapid development in recent years, also thanks to AI, the pace of increase in life expectancy will only go up. Since 2000, the increase in life expectancy has not improved the average quality of life of humans, it has led to a large population of old people specifically in the developed world.

We tried to provide free medical facilities to our population but with the increase in average life expectancy, more and more of these resources are getting cornered by the senior citizens, and resources left for the next generation are depleting at an alarming rate. Most of the sinking funds and medical funds are deep in red. This means, the younger generations are funding these funds today based on trust but when they will actually need it the funds will go bankrupt and will not be able to honour the commitments.

Finance is one thing, but the biggest problem higher life expectancy created was an increase in the average age of conceiving offspring. The average age of first-time mothers in the United States has increased to 27 from 24 in 2000. It is soon expected to cross 30. This is alarming as delayed pregnancy leads to deterioration in the genes of the next generation. Delay in the decision to bear a child is also leading to a lower number of childbirths and hence increasing the average age of the population."

Cyrus realized that people were not following what he was saying. He is losing them. Cyrus realizes that it would be foolish on his part to expect the general public to understand the irony of higher life expectancy leading to the extinction of mankind. Even Micky and Leo took some time to understand the irony. He mentally made a note to reserve this discussion for debates with intellectuals. Many such media debates have been planned by the project team over the next few days.

Cyrus considered getting into technical details of how the gene editing was achieved and what are the expected benefits of the project, but he realized that it would be too heavy for the crowd that is in front of him. Anyway, they are tired after

3 long speeches. Not only tired but confused, frustrated, scared, agitated, all at the same time. Cyrus decides to directly move to Q&A, that way he would be able to hold their attention and would be able to focus on things they want to hear.

As soon as Cyrus opened the platform for questions, they were a deluge of questions. The management team of the event was smart enough to anticipate it in advance. They were well prepared. They got the moderators in place; microphones were placed in different sections of the crowd so that they could ask questions from where they were and it would be audible to all.

People started asking questions one by one and Cyrus and Leo patiently kept answering each one of them.

1. *What if somebody turned 60 on 5th Jan 2025, will he die on 5th Jan 2030 or 20th January 2030? Because, if he dies on 5th Jan 2030, he will get 15 days less than 5 years?*
2. *What if somebody has not taken any Covid vaccine? Will he be saved as gene editing was not done in his case?*
3. *What if Russia and China are able to destroy the Satellites of SatelliteLink? Will the project turn off if the satellites are destroyed?*
4. *If I hide in a cave, will I be able to avoid the rays from the Satellite and save myself?*
5. *What are the side effects of gene editing and satellite rays?*
6. *What kind of compensation government will pay us for killing us?*

7. *What facilities would be available on Mars in 2030? Right now, I don't think anybody can survive on Mars. Or have we already colonized parts of Mars?*
8. *Can I go to Mars before I turn 65 and return back when I turn 66? Will I be saved in that case?*
9. *If you send all the old people to Mars, and if you think old people are a problem for Mother Earth, are you not killing the project Mars and the objective behind it by sending old people there? Will we look for another planet to create the backup?*
10. *Do we have enough seats on the flight to Mars? Do we need to make the booking early? I hope there won't be any discrimination in seat allocation.*
11. *Is the fight to Mars truly free or there are some hidden costs?*

The questions and answers have been continuing for more than two hours now. Cyrus and Leo patiently answered each question. The crowd was initially attentively listening to each and every question and answer. But with passing time the questions started becoming more repetitive and bizarre. The crowd started losing interest. People have started drifting out of the stadium, the crowd is thinning with each passing minute. It is not showing on the faces of Cyrus and Leo but they are taking a sigh of relief. They were worried that people would become very violent and it would become a riot-like situation. Thankfully that has been avoided. But inside their minds, they are well aware that the battle has just begun. Due to the time difference, people in the eastern hemisphere are just waking up to the news. Countries like China, Russia, North Korea, and Iran would not take this well. They need to move out of the stadium to handle the bigger battles.

Chapter 2:
The World Reacts

China

The coronation ceremony of the 51st President of the United States of America is over. But nobody is talking about the coronation. All the media channels are talking about is only one thing – Project Auto-off @ 65!

The chatter about the project is not limited to TV channels or limited to Washington. The project is the most talked about item on all social media channels and TV channels across the globe. The talk is not limited to media and the internet, the project is the most talked about topic in every drawing room, every office canteen, and every nook and corner.

The announcement has taken the whole world by surprise and as expected the feedback is not good. People across the world are angry.

China, in recent times, has become the second biggest superpower in the world and is giving the United States of America a close fight. China historically has suffered from the problem of a large population. This country, similar to India, believed that its rapidly rising population is compromising its development. Hence to solve the problem of the rising population, China implemented the one-child policy way back in 1979. The one-child policy of China has helped it control its rapidly rising population and enabled it to become a developed country. India on the other hand, which faced a similar population problem, did not come up with any such policy and is still struggling with a rapidly rising population. As a result, it is still a developing country.

China capitalized on the demographic dividend over the last 4 decades and managed to gather the 'escape velocity' to escape from poverty to developed nation status.

However, the situation has changed in the last few years. Due to the one-child policy over the last four decades, the number of births in China reduced significantly. The impact of the fall in birth rate is now being felt in the country with a decreasing number of youngsters joining the workforce every year. This problem coupled with the problem of the rising greying population due to increased life expectancy is leading to a problem of imbalance in the labour market, economy, and society.

Young Chinese citizens joining the workforce face the problem of having too many dependents. As every Chinese citizen born in the last four decades a single child, he or she has the responsibility of both parents. The same is true for his or her life partner. This means each couple has four ageing parents dependent on them. Apart from that, they also have dependent children. On top of this, due to the increase in average life expectancy and thanks to modern medicine, many youngsters have the added responsibility of looking after their grandparents. On average, a typical 30-year-old Chinese man will have two kids (now China has reversed the one-child policy), two parents, two parents of the spouse, and eight grandparents to look after. On one couple there would be financial, medical, and other responsibilities of sixteen people including themselves. With so many dependents it is difficult for both husband and wife to work as they have to look after young kids, ailing parents, and bed-ridden grandparents. This means the total responsibility of sixteen family members falls on the shoulders of one earning member in the family. This has become unsustainable and hence forced the Chinese government to reverse the one-child policy.

Now they have moved to the opposite end of the spectrum where they are incentivising citizens to have more children. They are giving financial and other incentives. However, any child born today will join the workforce after 20 years. Therefore, these incentives to have babies will not solve the problem of too many dependents in China in the near future. It will only become more aggravated as average life expectancy keeps on increasing, again thanks to modern medicine. They need to find a quick fix to their problem and Project Auto-off @ 65 is an effective quick fix. The moral dilemma of getting rid of dependent parents and grandparents is a different issue.

Mr. Lee is a typical citizen of China. He is 35 years old and has two daughters who are 5 years and 2 years old. His mother and father are 60 and 65 years old. His wife's parents are also of similar age. His grandparents are still alive and they are 86 and 89 years old. His grandparents from his mother's side are also alive and they are around 83 and 86 years old. All his parents and grandparents are dependent on him. Similarly, her wife's parents and grandparents are also dependent on him. His wife is a homemaker. Including himself, he has 16 dependents. He has a decent job by China's standards, but he is struggling financially. With his one income, he needs to take care of 16 people. Most of his grandparents are perennially ill. The medical bills of senior citizens eat up more than fifty percent of his income. He has bought a large house so that all 16 members of his family can stay together. Earlier his grandparents used to live separately, but now with deteriorating health, it is not possible to leave them alone. His wife is educated and wants to work but is not able to work as she has to take care of two young children and 12 senior citizens, eight of which are constantly in and out of the hospital.

Mr. Lee was forced to take a loan to buy the house. Now he is struggling to pay the EMIs. He has other EMIs for car, education, and marriage which he needs to pay. He is borrowing at higher rates of interest to pay back earlier loans. He is not even able to make payments on his credit card dues in time. He is in a debt trap and does not know how to come out of it. Due to his financial troubles, he has gone into depression. The regular visits to the hospital for their grandparents are not helping either. He has considered committing suicide several times. He is worried about his kids and wants to give them a better future. He does not want to leave them under the burden of debt.

The announcement of the project Auto-off @ 65 cured him of his depression. His suicidal thoughts vanished. He can now see light at the end of the tunnel. He now has hope of getting out of the debt trap. He would have preferred if the conspirators had not delayed the D-Day by 5 years.

Sitting in his drawing room he is watching the news and discussing with his wife. They are avoiding directly supporting the project as they would appear as bad humans and ungrateful offsprings. But both of them know what the other feels. Both of them are bogged down by the weight of so many dependents. They want their independence back and this project will give them back their independence soon.

Mr. Jia Lian, the current President of the Republic of China is 71 years old. He is the general secretary of the Chinese Communist Party (CCP) and the chairman of the Central Military Commission (CMC). He has been President of China since 2012. China is a non-democratic, communist country and Lian being the trusted leader of the CCP, has absolute power when it comes to politics in China. His word is the last word in China. As China is the second most powerful country in the world, Mr. Lian is the second most powerful man on Earth.

President Micky, Leo, and Cyrus are apprehensive about how China and Mr. Lian would react. Over the last few years, China has become an export powerhouse. The USA is one of the biggest importers from China and is dependent on it for many critical and non-critical supplies. China has been recording trade surpluses for several years while the USA's trade has been in deficit for several years. The USA's trade deficit with China is more than $350 billion which is bigger than the GDP of many countries. Due to the increasing trade deficit, USA has been borrowing from China and China has been investing in US sovereign bonds. Today, China holds more than $20 Trillion worth of US sovereign bonds.

Despite extensive trade, due to China's aspirations of becoming the biggest superpower in the world, the USA does not have very cordial relationships with China. In the past, China has threatened to destroy the US dollar's hegemony in world trade by offloading trillions worth of US sovereign bonds in the global market. The USA cannot afford any such move by China. They need China for their imports and they need China also to buy their sovereign bonds.

China is also dependent on the USA for its exports. China managed to become a developed country by building humongous factories that are now factories of the world. This has helped China create a lot of jobs and pull out their people from the clutches of poverty. China cannot afford to switch off these factories and the USA is their biggest customer. USA and China's situation is like that of marriage, where both partners regularly quarrel but divorce is not an option.

As expected, China was the first to react. Mr. Lian, in deviance from the standard practice, went live on national television and openly threatened the USA for the **'Act of War'**. China termed Project Auto-off @ 65 as an act of war as it effectively means killing millions of Chinese citizens. China gave an ultimatum to the USA to reverse the project and threatened that otherwise China would be left with no option but to nuke all US facilities involved in the project including SatelliteLink's thousands of satellites in outer space. Chinese government issues several threats during the public address and immediately after the public address by the President. All these threats were concerning but one that specifically stood out was the threat to shut down Nikla car gigafactory in Shanghai. Nikla is a trillion-dollar electric car company owned by Leo which has their biggest manufacturing base in Shanghai.

Micky, Leo, and Cyrus were concerned about China and their concern has come true. The USA cannot afford to get into war with China at this stage, or any stage. They were betting that China with its struggles with the problem of the rising number of dependents would welcome this project. They did not expect China to publicly welcome the project. However, they expected them to do some public sloganeering just to keep the local population happy, but eventually support the project. During COVID-19, there was unconfirmed news that China was mercilessly killing citizens having COVID-19 to control the spread of COVID-19. China has a reputation for being merciless and moral-less when it comes to meeting national development and other goals. China would be one of the biggest beneficiaries of Project Auto-off @ 65. Leo was expecting that Mr. Lian would do public posturing but would support the project. Leo was aware that Mr. Lian may oppose the project due to personal reasons as Mr. Lian will also have to shut off from the Earth in 5 years. Leo has specifically planned a base on Mars reserved for Mr. Lian and his associates to keep him in good humour.

India

India is the most populated nation in the world with a population of 1.43 billion people. The average life expectancy in India is 70 years. India is still a developing country and per capita income is only $2500. If one has to identify the biggest reason why India is still a developing country and not a developed country – the answer is population.

In 1979, before China introduced the one-child policy, China's population was around 968 million and India's population was around 683 million. China was roughly 1.4 times the size of India in terms of population. India's per capita income was around $580 which was almost twice that of China at $307. Today India's population is 1.43 billion, more than 2 times it was in 1979, while China's population is 1.41 billion people. India's per capita income is $2500 while China's per capita income is around $13,000, more than 5 times. Many things have worked for China in the last 45 years, population control is one of the major factors. The simple answer to the question – what is the one main reason why China has managed to become a developed nation while India is still a developing nation, although China was substantially behind India both in per capita income and population control in 1979, the answer is – population control via one-child policy.

The median age in India today is 28 years. Only 7.2% of the population is above 65 years of age. Considering India's population is 1.43 billion, roughly 100 million people are above the Auto-off age. Adding 4% population who are between 60 and 65 today, by 2030, around 160 million people will be past the cut-off age. This means, that on January 1, 2030, 160 million people will die due to project Auto-off @ 65. To put things in perspective, the population of the USA is around 330 million. This means a number equal to half the population of the USA will die in India alone.

What it means is that population is the biggest problem holding back India and it will substantially benefit if the non-productive population of the country is eliminated. However, the population is so large that the number of people who would be eliminated is also very large, and killing so many people at one go is no less than a massacre. It will easily qualify to be the biggest massacre in the history of mankind. But numbers do not give the real picture. The real picture is when you sit in a drawing room of an Indian joint family and understand their perspective on health, old age, medical, burden on the next generation, and willingness to live long – basically on Project Auto-off @ 65.

Laxman is a typical 45-year-old Indian male. He is sitting in his drawing room and following the coronation of the President of the USA, live on TV. Laxman is around 6 feet tall, is thin and has an athletic frame. He is good looking but his hair is receding at a fast pace. He is very conscious of receding hairline; he understands this is a sign of him getting old. He does not want to grow old. A couple of years ago, during a routine medical test, he came to know that he has multiple lifestyle diseases including thyroid, sugar, and cholesterol. Thankfully it was a borderline case. All these parameters have just crossed the border and Laxman still had time to control them.

Laxman took immediate action, controlled his diet, planned his lifestyle, and got into active sports, two hours every day. This helped him get the parameters back within the border. However, the threat on the border still looms and Laxman is very conscious of it. He does not want to lower his guard. Laxman fears getting ill, bedridden, and dependent more than he fears death. He wants to live a good healthy life all his life. He knows that old age and illness are as much a certainty in life as much as death. He knows that everybody will grow old and before death everybody has to go through the cycle of old age, illness, dependence, and death. Unless of course, somebody dies in an accident or sudden heart attack, etc. He understands that sudden death is also bad as it has a deep impact on the family which is not prepared.

Laxman is a typical well-educated Indian who is doing well in his career, running his own business, has a good lifestyle and a happy married life. His wife is also well-educated like him and works for a software company. She is also doing very well in her career. They have two children; the elder son is 14 years old and is in 10th standard and the younger daughter is 11 years old and is in 6th standard. Both the kids are doing well in school, and are well-mannered, happy kids going to nearby convent school. Laxman belongs to the Marwari community where community ties are very strong and they have a large family, many living together, who are not living together are still well connected.

Laxman lives in Mumbai with his wife and children while his parents stay in a smaller town Bokaro. Laxman has three other siblings, two brothers and one sister, all living in different cities. Laxman's father and mother are 76 and 69-year-old respectively. His father has had many medical episodes in the last few years and has grown very feeble. But thankfully he is still doing fine and not dependent on anybody. His mother is still very fit and takes care of her household all by herself. Laxman's parents live all by themselves as all their children have migrated to larger cities for better employment opportunities. They have been living alone for more than 20 years now. The father has retired long back and is now associated with a small CA firm from where he earns enough to maintain their simple lifestyle. Laxman's mother was always a homemaker and is very good at that.

His wife Urmila's parents live in another small town Bargarh. Her father runs a decent-sized CA firm, which he has built over the years and where his son joined 25 years back. Urmila's father is now 85 years old, still decently fit and independent. However, with a few old age-related lifestyle diseases that have restricted his lifestyle. He avoids venturing out of his house much. His son has taken over the firm and running it very well. He goes to his office more to spend time and to not feel like a retired person. Urmila's mother who is a homemaker, is now 75 years old and has some hearing and visibility problems. However, she is also decently fit and independent. They are maintaining their good lifestyle along with their son's family. They live in a joint family of 6 people.

Laxman just heard the announcement made by the President. It is still early in the morning in India, around 5 am. He woke up just to catch the President's coronation live, Urmila was still fast asleep. Laxman shouted Urmila's name asking her to come to the drawing room. When Urmila did not react, he shouted three or four times again and again. He ends up waking up the whole family with his shouting. Urmila is visibly and vocally irritated and so are the children. He ignored their irritation without even considering it. He immediately started briefing them on what just happened. He started – **"we all will die at 65."**

Laxman was expecting them to get shocked. But none of them even bat an eyelid. Maybe because they are still half asleep or maybe because they don't care to live beyond 65 or maybe the concept of 65 is too far in the future to care. Urmila is even more irritated now and Laxman is now defensive. He needs to do something now otherwise Urmila is so furious that she might hit him with his squash racquet lying next to him.

Laxman hurriedly blurted out – "your parents will die on 1st January 2030." Now Urmila is shocked. She is completely awake, but even more irritated and furious. She thinks this is some kind of a joke and now she is furious enough to beat him. Laxman realized his mistake and immediately added - "and so will mine."

Now Urmila is taken aback and confused. She just dropped herself on the sofa, half intentionally half out of shock. She is no longer furious but very confused. The TV is still playing in the background and the commentators continue to discuss something with heightened alarm. She stares at the TV, then she stares back at Laxman expecting him to explain himself. By now the children have also forgotten their sleep, wide awake and settled on the sofa. Laxman went on continuously for the next 5 minutes explaining what he had learned about the project Auto-off @ 65. All 3 of them listened to Laxman without interrupting. At the end of Laxman's monologue, they were shocked, shaken, and more taken aback. They had a lot of questions and nobody was able to control himself or herself. Urmila, as usual, shut down the kids with a single wave of her hands and started off with her questions.

First, it was more frustration than questions. *"How can they do it? Who is the President of the USA to decide how long my parents will live? All the people behind this project should be lined up and shot bullets in their heads."* Now the reality is slowing sinking into her and she asks – *"Is there a way to stop this? My parents are still very healthy, why would they kill healthy people? – they should target only bedridden people."*

Laxman spent another 5 minutes explaining that Leo and Cyrus have clearly mentioned that there is no exception and there is no stopping or reversing this. He also tried to explain 'save mankind reason' which he himself has not been able to digest yet. He did not talk about the Mars exit route as he believed that it was not a real option for his old parents.

Urmila is also a 45-year-old, healthy, well-educated, well-to-do independent woman. She is also an out-and-out feminist. Maybe if Mickey, Leo, and Cyrus had kept one female as their co-conspirator it would have been easier to convince Urmila. Urmila is not too worried about herself or her kids. For her 65 is too far in the future. She is worried about her parents.

She is a workaholic. She at times gets frustrated with her round-the-clock work hours and talks about taking an early retirement and enjoying life. However, she is not serious about it. She loves to travel, party, chill, and watch OTT and loves it when she gets an extra holiday. She romanticizes the idea of taking an early retirement and doing this every day, but in her heart, she knows she can't do it every day. Travelling for a few days, once in a while, is good enough for her, she is not a travel junkie. She would rather lie down in bed and watch TV, but she already feels she has run out of OTT content, she can't imagine doing it for days, months, and years. Retirement means doing this every day for years. Similarly, she can't spend too much time partying or socializing with friends and family. She likes small doses of all these but she cannot do it every day. Currently, she cannot picture life beyond 60. Doing all these for 5 years, getting bored of it, and dying at 65 is not that bad a thing for her. At least now, she does not have a lust for a very long life. She wants her kids to complete their education, get married, get settled, and go on with their life. She also does not have any plans to live with her kids in old age and would prefer they live their own independent lives as she has lived all her life.

Being an independent woman, a feminist, she also does not want to be reliant on anybody during her old age. At least today, she would prefer to die before the start of the old age problem. On giving it a thought, she thinks 5 years of retirement is a bit too short. She would have preferred 10 years. But for her 10 is good enough. She would be fine if the conspirators revised the project to Auto-off @ 70.

She thinks her parents are fit enough to last another 10 years. She wants 10 more years for them. After a lot of debate and discussion, she concluded – *"If the conspirators revise the project and add 5 years for everybody, I would be fine with it. I would support it. I see the merit of not going through old age and the merit of leaving the resources to our next generation."*

Laxman also agreed with her. He also has a very similar view of the world. For his father he believes even 5 years should be enough as he is already very feeble, he would prefer he does not have to suffer too much in old age. His mother, he thinks is very fit and can have a good disease-free life for another 10 years. He agrees with his wife's idea of another 10 years instead of 5 for parents and Auto-off at 70 for themselves.

Laxman and Urmila went on to discuss the impact of the project on their various friends and family. The kids also participated in the discussion. Kids seemed okay with the current cut-off also for them even 65 is too far in the future and they are not as attached to their grandparents as their parents are. For them, five years is also a long time.

After a couple of hours of discussion, both Laxman and Urmila were surprised that they were not as angry as they were when they heard about the idea. They now see merit in the idea not only at the country level but also at the individual level and their personal level.

Laxman turned to the TV and increased the volume. A heated debate is going on, on the TV. Most of the people on the TV are politicians and most of them are very angry. There are few intellectuals who are logically debating the merits and demerits of the project. They are highlighting and admitting there are certain merits to the plan but they are also taking a stand against the project but in a much milder fashion. There are few representatives of certain religious groups, they are mostly talking nonsense. They are not even trying to use any logic and each time their argument is that humans should not act like Gods and should not go against God's plan. Few went on to declare that conspirators of the plan should be hanged and threatened to issue *Fatwah* in their name.

Politicians of the ruling party are also speaking against the plan and showing helplessness as it was hatched by the USA without their knowledge. The opposition politicians have sensed an opportunity. They are against the plan and they know public sentiment is against the plan. They want to use this to corner the government of the day. They are asking questions from the ruling party; they are making fun of the ruling party for being so helpless and not being able to do anything. The Prime Minister of India has carefully crafted his image as that of a strongman, with a 56-inch chest, this helplessness goes against his image. They also question the Prime Minister of the country as to whether he is one of the co-conspirators. Micky in his speech mentioned that many world leaders were in the know of the plan. The PM of India is considered to be a very close friend of the President of the USA. They have been very friendly since the first term of both the leaders, and have never shielded from making a public display of their friendship, at times to garner more votes.

Mr. Mehta, the Prime Minister of India has an image of a strong nationalist leader. He has been in the post of Prime Minister of India for more than 10 years now and is one of the longest-serving leaders in any democratic country currently. He is the only democratically elected leader to survive Covid.

During his reign, the stature of India has risen rapidly. Today he is considered to be one of the most powerful world leaders along with Mr. Micky Cash of the USA, Mr. Lian of China, and Mr. Turin of Russia. But unlike other leaders mentioned above he is considered a powerful leader not because of the military power or economic power but because of how he handles himself and his vision for the world. During the G20 Presidency, India, Under the leadership of Mr. Mehta, gave the world vision of *"One Earth, One Family, One Future"*. Today, the world is looking up to the PM of India for peace mediation among hostile countries, to broker peace between Ukraine and USSR and Israel and Iran respectively. Mr. Mehta is considered to be one of the most balanced and mature world leaders today. The world has started looking up to India and Mr. Mehta to lead the way.

However, he is not short of critics. His distractors like to paint his image as a Hindu fundamentalist leader who is against the Muslim community. They call him names like 'butcher of Gujarat'. Many blame him for the massacre of the Muslim community during the Gujarat riots in 2004 when he was the Chief Minister of Gujarat. 20 years have passed since then and the highest court of India, the Supreme Court, has absolved him from the blame, still, this black spot on the image of Mr. Mehta fails to disappear. He would not like his name getting associated with another massacre now.

Mr. Mehta is also known for his ability to take strong steps in the long-term interest of people even if it leads to short-term pains. Mr. Mehta has been acclaimed and criticized, at the same time, by the public, media, and experts for his drastic step of demonetization in 2016. Demonetization led to a lot of trouble for people but in the long term, it is believed to have laid the foundation for India's robust fintech economy, something in which India is a role model for many others. India's UPI payment system is considered to be the most advanced in the world not only in terms of technology but also in terms of grassroots penetration. Many countries are partnering with India to launch the technology in their country.

In March 2020, Mr. Mehta announced and executed the world's strictest Covid lockdown. It was the harshest lockdown anywhere in the world. People were taken by surprise. Many people went through a lot of suffering specially the daily wage earner. Many lives were lost, especially of daily wage earners who tried to migrate back to their villages due to lack of work. However, in the long run, it is believed that the strict lockdown helped in saving millions of lives.

The opposition leaders are making full use of this opportunity. They are questioning Mr. Mehta's strongman credentials and his friendship with the evil President of the United States of America. They are questioning his chest size, they are calling him a co-conspirator, and they are calling him a butcher. Above all, they are questioning his absence in this crucial moment. Three hours have passed since the speech of the President of the USA. Although it is still early hours in India, there is already a lot of anger among people. There is a lot of anger in the newsrooms, it appears that newsrooms are egging up anger among people. Already there are unconfirmed reports of riot-like situations in various parts of India. People are angry and the anger is now boiling down to Indian streets. Something needs to be done and needs to be done fast.

Mr. Mehta is marked by his absence. It appears that home ministry has sprung into action and ensuring that law & order is maintained. Even the Indian army has been put on high alert. Central Reserve Police Force (CRPF), India's primary agency entrusted to maintain law and order has been deployed in sensitive areas.

Suddenly all news channels started flashing that the Prime Minister of India would address the nation at noon India time. There is still 4 hours to go before noon. Laxman saw the news flash and wondered why after 4 hours. Why is he not addressing the people now? What is he doing which is more important than this?

Suddenly the tone and direction of the news channel has shifted. Now everybody is discussing what Mr. Mehta might say. Experts are trying to make intelligent guesses; the opposition is questioning the delay. Some are also wondering if there is a high-level video conference happening between Mr. Mehta and leaders of the world especially the President of the USA, which is holding Mr. Mehta back.

Laxman and Urmila, like the rest of the world, kept themselves glued to the TV screens for the next few hours waiting with bated breath on how things would unfold and what Mr. Mehta would say in his address to the nation. They have decided to not send their children to school today as there is fear of untoward incidents on the street. They themselves have decided to not go to the office today and spend the rest of the day in front of the TV following the latest developments. Since then, the government of India has announced a national *bandh* (shutdown/holiday) to avoid the law-and-order situation getting out of control. They have also imposed restrictions on meetings or assemblies of more than 5 people to avoid riots. Social media channels like WhatsApp, Facebook, and Y have been instructed to ensure no false rumours circulate on the platform which can lead to a law-and-order situation. They have been warned that they will be shut down if any such incident comes to the notice of the administration.

It's noon and exactly in time Mr. Mehta is on screen across news channels. There is not a single news channel in the country that is covering anything else, despite the rapidly evolving situation and various developments across the world due to this project. Mr. Mehta begins his address to the Public.

"Friends, brothers and sisters, my fellow citizens of India. Like all of you, I also heard about the Project Auto-off @ 65 announced by the President of the United States of America today morning.

I share your anger about what they have done and we have already communicated our displeasure to the USA government. We are in touch with various world leaders from across the world and we will keep you informed about the development. With the limited discussion that we have had, it appears that the project is already been activated. There is no way to stop the project and there is no way of going back. However, we are still working on the possibility of somehow switching off this project. Please note, that the first switch-off under this project is planned for 1st January 2030. Hence, we have 5 years before any negative impact of this project is felt. We will keep working with our scientists, technologists, and other nations to find ways to stop this project. We request all of you to not make any hasty decision. Do not react. We have time. We have solved many big problems in the past, we have recently successfully handled COVID-19 and come out of it stronger, and we will do the same with this crisis. Please trust us and maintain calm. Do not give in to fearmongers. Do not give in to rumours. We have time, please give us time and trust us. As your humble servant, I promise you, I will die before any of you die,"

Laxman is nodding his head. Yes, we have time. Yes, we can find some solution.

"I as Prime Minister of India accept that there has been intelligence failure at our end. USA, President Micky, Leo and Cyrus had been working on this for years and we did not get any whiff of it. But do appreciate that this project was a closely guarded project and even the US government was not aware of it. The outgoing President of the USA and his team had been caught by surprise as much as we had been astonished. Best-known intelligence agencies across the world including China, the USSR, the UK, and Israel were shocked as well. Hence, we cannot completely blame our intelligence agencies only,"

Now Laxman is not agreeing. How can there be such a colossal intelligence failure? Somebody needs to pay the price for this. Laxman finds it uncharacteristic for Mr. Mehta to admit colossal intelligence failure. He could have avoided speaking about it. Maybe he spoke about it because he wanted to indicate that he is not one of those world leaders who knew about the project as hinted by Micky. Laxman wondered why he did not mention that directly.

"I am 74 years old. If we are not able to stop the Project by 2029, I will die on 1st January 2030. I will die before any of you will die. As your humble servant, I promise you, I will die before any of you die,"

Laxman is mumbling, we want our parents to live, we don't want you to die.

"Mr. Mickey Cash, President of the United States of America, Mr. Leo Smith and Mr. Cyrus Clark are not evil people. All three of them have the interest of mankind at the top of their minds. I had interacted with them very closely during the Covid crisis. If it was not for the leadership of Mr. Cyrus Clark we might not have invented the Covid vaccination in time and might have lost many more lives.

They are not foolish people, they are very intelligent people, they are where they are today because they are intelligent. I have interacted with them many times and they are one of the most exceptional people I have interacted with in my life.

They are not only intelligent, they are geniuses. They are not only not evil, they have dedicated their lives for the betterment of mankind,"

Laxman is thinking – what is Mr. Mehta doing, where is he going? He is praising the evil trio which is looking to massacre millions, if not billions, of people.

"We have time, let's spend some time and understand what is the motive of the creators of the project. India's average age is much lower than USA or other developed countries. As a percentage of the population, more people will die due to this project in the USA than in India. Which means they have more to lose than us.

Both Mr. Micky Cash and Mr. Cyrus Clark are pretty old themselves and will be the first ones to switch off under the project. We need to appreciate that they have treated everybody equally including themselves. At least, they have not discriminated against us.

We also need to appreciate that they have given us time of 5 years. They have given us time to plan and get our house in order. The project will not disturb anything for the next 5 years. The switch-off under the project will start only in 2030.

We also need to agree with them that the rising average age of the human population is a challenge. The falling birth rate of humans is a challenge. We have to agree with them that elongating the old age of humans, by using medicines is not always elongating the happy days, most times it is just elongating the days of pain and suffering.

We also need to agree with them that due to the increase in human life, and the elongation of unproductive years of human life, we are stealing resources from our younger generation and future generations and spending on the old population. The old people are past their productive years, from the point of view of mankind, if we think without blinding our eyes with emotion, we should be spending more resources on the productive younger population. We should have a higher education budget and a lower medical budget.

We need to agree with them that if old people are there, they will occupy all the important positions and the younger generation will not get a chance. I have always maintained that even politicians should retire at 75 so that the younger generation gets a chance. In my view 75 should be the cut-off age for retirement, these three intelligent people have designed 65 years in the project. We might disagree with them on what is the right time. But we need to admit that the idea is correct. We can disagree with them that killing them is not the solution, but we have to agree with them that yes, we need to make space for the new and future generations.

According to Hindu texts, there are four stages of life in a human's life called the ashramas:

- ***Brahmacharya**: The first stage, when one is a student and lives with a guru*
- ***Grihastha**: The second stage, when one is a householder and married*
- ***Vanaprastha**: The third stage, when one gives up worldly life and goes into the forest to meditate*
- ***Sannyasa**: The final stage, when one renounces worldly ties and becomes an ascetic*

As per our ancient texts, up to the age of 25, we should study and gain knowledge. Between the ages of 25 and 50, we should look after our family. We should have kids, and contribute to the continuation of mankind and the well-being of our family. After 50, we should start abdicating our power, and wealth to the next generation. we should contribute to society. We should share our knowledge, wisdom and experience with the younger generation to help them grow and prosper and help mankind prosper.

However, at the age of 75, we should abandon the world and not use any further resources of the world. We should leave it for future generations. If we keep elongating our lives, our future generation will grow in shortage of resources. Due to the shortage of resources, they will grow weak and weak people will give birth to weaker offspring. Hence, in the interest of our future generation and mankind, we need to move out, we need to switch off,"

Laxman is surprised and shocked. He was not expecting this. Mr. Mehta has just actually wholeheartedly supported the project. Is he one of the co-conspirators?

"However, I will never support killing people. We cannot kill our own people; we cannot kill human beings. I believe humans should themselves, on their own will, take up vanvas or move to the jungle. I really appreciate Mr. Leo Smith's offer to provide free transportation to Mars. Mars is the new Vanvas of our times. If our experienced, knowledgeable, people move to Mars, we will be able to colonize Mars faster and would be able to create a second home for humankind.

I will talk to Mr. President of USA and find a solution on how we can reconfigure the project to 75 years instead of 65 years. I will work with Mr. Leo Smith to ensure that transportation to Mars and staying on Mars is as comfortable and safe as possible.

Friends, before I end, I would like to request you. Please have patience, we have time. Nothing is going to happen immediately, we have 5 years before 1 Jan 2030, when the first switch-off is planned. If required we will fight with USA, but we will ensure that the project is reconfigured to 75 years from the current 65 years."

Laxman can now see through the speech of Mr. Mehta. He has not only wholeheartedly endorsed the project Auto-off @ 65 but also justified it using Hindu Texts. He repeated regularly that we have time so that people do not take impulsive action and law and order can be maintained. He talked about increasing the switch-off age from 65 to 75 to move the focus from death to timing. That was just playing to the gallery of voters. I believe he also knows that no change is possible now. The masterstroke was when he equated switch off / transportation to Mars to *Vanvas*.

Mr. Mehta has played a very good balancing act of pacifying his people, his voters, and fellow countrymen but at the same time, he has openly endorsed the project. He again has lived up to his image of a visionary leader who is willing to make strong decisions for long-term vision even if it means short-term pains.

Stop! He is 74, and next year he is turning 75. Did he just hint that he will be abdicating the post of Prime Minister next year and taking *Sannyasa*?

Russia

The Union of Soviet Socialist Republic (USSR) has been the biggest competitor of the USA for several decades. Cold War between the two countries lasted for more than 40 years from the end of World War II until the fall of the Soviet Union in 1991.

During the Cold War, the world was a bipolar world. Unlike today, there was not one superpower, but two. There were supporters of the USA or supporters of Russia. There were very few countries that were not aligned. The North Atlantic Treaty Organization (NATO) created in 1949 was also the result of the Cold War between the USA and USSR. NATO was created in 1949 by the United States, Canada, and several Western European nations to provide collective security against the Soviet Union. NATO was in effect, an association of countries supporting the USA trying to keep USSR in check. NATO's one of most important stated goals was – "*Ensure that an attack on one nation is an attack on all.*"

With the fall of the Soviet Union in 1991, the USA became the only superpower in the world. With the rise of China in the last two decades, it is now very close to becoming the second superpower in current times. However, Russia also wants to regain its erstwhile glory. Russia and the USA still consider each other as an enemy nation and always bickering with each other.

Ukraine war has deteriorated the relationship between the erstwhile foes, the USA and Russia even further. Russia's attack and ingression of Ukraine has brought the USSR and USA again on the edge of war. Ukraine is not a NATO country and hence USA has not entered the war to support Ukraine but many NATO countries in the proximity of Ukraine are threatened. USA has resisted entering the war overtly as that could lead to the expansion of war and ultimately lead to World War III. However, the USA is covertly supporting Ukraine with weapons and money. The USA has also announced several sanctions against Russia, financial and other sanctions to make Russia weak. Several NATO countries, partners of the USA are also overtly or covertly supporting Ukraine. Russia is not happy about it and will leave no opportunity to get back at the USA.

Mr. Turin, aged 72 years is the President of Russia. Mr. Turin portrays Russia as a thriving democracy by holding periodic elections/referendums on his own Presidentship but for all practical purposes is an authoritarian dictatorship. He has been ruling Russia since 1999. He was designated Prime Minister from 1999 to 2000 and between 2008 and 2012. He held the designation of President between 2000 to 2008. Now again since 2012, he is holding the designation of President. As there is a constitutional restriction on any person holding the post of President for more than 2 continuous terms, he took up the designation of Prime Minister between 2008 and 2012. From 2012 he had been holding the post of President again for two terms of 6 years each, till 2024. In 2024 he was supposed to demit office, however, he got a constitutional amendment passed in 2020, which in effect excluded his previous stint from calculation and allowed him to continue as President even beyond 2024.

By whatever name called, he has been the ultimate supreme autocratic leader of Russia since 1999. Even after being in power for 25 years, his hunger for power has not been satisfied. He also aspires to develop Russia back to its stature as a superpower. He wants to unite the erstwhile countries of the Soviet Republic and the invasion of Ukraine is part of this plan. Previously Mr. Turin annexed Crimea in 2014. Mr. Turin is a strong aggressive autocratic leader who does not shy away from war.

Russia is frustrated with the USA's and NATO's support of Ukraine in the war. Russia would not like to get into an all-out war with the USA and NATO countries. Together they are very powerful and resourceful. Russia also does not have powerful alliances like it used to have in the Cold War era however, China, and North Korea are its alliance partners. Partnership with China, the rising superpower, makes Russia very powerful and the USA and NATO countries also do not want to get into a war. Over the years the NATO countries, especially the prospering European Countries and their citizens have realised the futility of war. The USA is also slowly and steadily realising the futility of war and has been avoiding war especially when it comes to sending soldiers on the ground. The USA is fine with bombarding from top a country like Afghanistan with limited resources to strike back, but it is not in a financial or mental frame to get into war with erstwhile superpower Russia or upcoming superpower China. All these countries are nuclear power and all sides realise that it would be very difficult to contain the war if war broke out between these countries.

Hence, the only option left with Russia to get back at the USA is by breaking up its alliances and causing financial damage to its economy. In today's nuclear world, the superpower status is linked with financial muscle more as nuclear weapons have levelled the playing field and made military supremacy almost impossible. Nuclear weapons are so destructive that even a superpower cannot afford to get into war with another nuclear power.

Russia was confident that the USA would not fight alongside Ukraine as the USA and its public were in no mood to get into a war with a nuclear power. Russia believed its annexation of Ukraine would be as simple as Crimea in 2014. However, although the USA did not enter into a direct war with Russia, it has tried to cripple Russia and its economy with various sanctions including financial sanctions, technological sanctions and trade sanctions.

The announcement of Project Auto-off @65 is Russia's opportunity to cripple the USA without entering into a direct war. The USA has most probably gone ahead with this project without taking into confidence the NATO countries. The average life expectancy in these developed European countries which form part of the NATO is quite high at 78.9 years. Auto-off @ 65 means killing people 14 years early. This project also means that the USA is planning or plotting to kill millions in NATO partner countries. Most of these NATO partner countries are democracies and people in powerful posts are people above the age of 60. Russia has a very good opportunity to escalate matters and turn the NATO countries against the USA. These NATO countries would not be in a position to support the USA on project Auto-off as they are also answerable to their own vote bank. Ideally, Russia would like to create a situation of the USA vs the rest of the world. China has already threatened to nuke the USA for unilaterally initiating this project. Most other countries will follow. Russia has a golden opportunity to isolate the USA.

Personally, Mr. Turin is 72 years old. Hence, will also be switched off on 1st January 2030. Mr. Turin maintains an image of a very healthy and medically fit person. He loves to show off his muscular body in shirtless pictures leaked to international media. Although there are unconfirmed reports that Mr. Turin is suffering from some serious disease. There are rumours that he is using his body double for public appearances. His recent pictures of him walking with a limp added fuel to the speculation. Mr. Turin would be genuinely pissed off with USA for initiating this project but at the same time, he would understand the reality that what is done is done. He would know that he only has 5 more years and would like to use this opportunity to corner the USA and fulfil his ambition to make Russia more powerful than the USA. Unless he desperately wants to live longer and wants to book his ticket on the Mars vehicle, he has nothing to lose from war-mongering against the USA.

The national TV of Russia announced a live broadcast of Mr. Turin's address to the nation and the world. The address is scheduled in 1 hour. The point to note in this one-line news flash is the words 'and the world'. This is a first. Whenever the President of Russia makes a live speech on national TV it is referred to as an address to the nation. For the first time, it is being referred to as an address to the world. This is a clear indication of what to expect. Russia is about to declare war with the USA or at least threaten a war. Nobody in their wildest of dreams can expect Mr. Turin to support the USA where he has a once-in-a-lifetime opportunity to isolate USA.

Mr. Turin is on the screen ready to deliver his speech. With the limited image where he is sitting on a chair and front facing the camera, it is difficult to conclude whether he is having any medical issues. His face looks a bit pale; he appears a bit weak but nothing alarming.

"Citizens of Russia and the world, today I am speaking to you at a moment when the existence of mankind has been put at risk by a few mad men from the USA. USA and the President of the USA have conspired to massacre millions and billions of people. They have the audacity of portraying the planned massacre as a project to 'save mankind' when actually, they are doing exactly the opposite.

The USA has been always the nation of madmen, trigger-happy, senile, madmen and this new president of the USA, Micky Cash is no less. I have seen many madmen occupy the chair of the president in the last 24 years, but Micky Cash is the craziest of them all.

Project Auto-off @ 65 is a conspiracy hacked by three billionaires to blackmail the world. They will blackmail millions of people across the world to buy tickets to Mars and then dump them in a God-forsaken place to die. They will dump millions of human beings in a place without oxygen, water, food or any other basic amenities required for humans to survive. How do we expect people above the age of 65 to survive an excruciating journey to space let alone the harsh environment of space? They will blackmail you first to buy a ticket to space. Then they will blackmail the rest of the population under the age of 65. They will blackmail them to buy return tickets from space for their parents.

Please do not believe them when they are saying that they themselves will get switched off on 1st January 2030. They are saying this just to fool you. They will have palaces built for themselves from the money they blackmail out of you and live even more luxurious lives than they are already living. There is no end to the greed of these mad billionaires.

These greedy billionaires are greedy for money and power. They want to suck you dry. They want to make slaves out of all of you. Leo Smith will insert chips in the brains of each one of us to control us. Our lives would be reduced to the life of a bot by these crazy tech billionaires.

We need to act and act now. We need to stop these mad billionaires from destroying the world and making us slaves. Russia is declaring war on the USA with immediate effect. Russia is giving the USA 7-day time to switch off project Auto-off @ 65. If the USA does not switch off the project within the next 7 days, we will not shy away from Nuking USA. We have enough nuclear power to destroy these crazy billionaires and their evil tech empires.

Any country that supports the USA will be considered an enemy of Russia. Russia will nuke these countries too. The NATO countries are given 3-day time to disassociate themselves from the USA. If the head of state of any of the NATO countries does not publicly announce disassociation with the USA within 3 days, it would be considered an enemy country by Russia and would be nuked. The USA has a population of around 350 million. If we nuke the USA and kill each and every American, then also we would be doing service to mankind. We would be saving around 700 million people, almost twice the population of the USA from being massacred on 1st January 2030.

Russia calls upon all countries to join us in our war against the USA and these mad billionaires who are controlling the USA. I have already spoken to China and North Korea. They have pledged support in our war against the USA. We call upon NATO countries, Australia, Canada, India, Saudi Arabia and all the other countries of the world to join us in this war.

We also call upon citizens of the USA to join us in war against these 3 madmen. We urge you to revolt against these billionaires who are trying to control your destinies. They call me an Autocrat, and they call President Lian of China an Autocrat, but did they take your permission before deciding to kill millions of people? If you do not revolt now, you deserve to die. We are giving 7 days to citizens of the USA to revolt against these madmen and bring them down. Any help you require for doing the same you can reach out to us. In 7 days if you do not revolt and bring down these billionaires, Russia will be forced to Nuke you in the interest of the world. USA's democracy was always a façade, in reality people of the US have no power. They are puppets in the hands of their politicians and businessmen. It is time for the people of the USA to wake up and take control of their lives.

We announce the following sanctions against the USA with immediate effect:

a. *USA would not be allowed to launch any satellite in space. These satellites are their core weapon in this project.*

b. *USA would not be allowed to sell any medicine outside the USA. They have gene-edited us using these medicines.*

c. *USA tech companies like Google, Facebook, Nvidia, Netflix and all major tech companies of the USA having turnover of more than $ 10 Billion are banned from operating in the rest of the world. They need to close their offices immediately otherwise these offices will be destroyed.*

d. *the USA is sanctioned from doing any trade with the rest of the world. Any country participating in trade with the USA would be considered an enemy nation and co-conspirator.*

e. *Sanctions are being announced against the use of the US dollar. Any country using the US dollar for any transaction would be in defiance of sanctions and would be considered an enemy nation.*

We again call upon leaders of all countries to join us in the war against the USA. We again call upon citizens of the USA to join us in the war against the President of the USA and his co-conspirators. We need to Act. We need to destroy these crazy, greedy billionaires and their evil empire immediately."

The speech was telecasted live across the world. Hearing the speech chill ran down the spine of everybody watching the speech live. Russia has just declared the start of World War III. The conspirators had announced doomsday in 5 years, but Russia has expedited it to 7 days. In 7 days, the world is going into a nuclear war!

Rest of the World

Islamic States

Iran

Iran or the Islamic Republic of Iran is a Muslim country and generally political opinion of the country is also governed by religious guidelines. Iran's population is around 91 million. The average life expectancy in Iran is around 75 years. Iran has one of the biggest oil and gas reserves in the world and has been part of many conflicts. Iran has an ongoing conflict with Iraq, another neighbouring country with large oil and gas reserves.

The United States of America supported Iraq during the Iran-Iraq war in the late nineties also called the First Gulf War. Since then, the USA has been a sworn enemy of Iran. The USA killed one of their very senior military leaders in a drone attack recently. The USA is also supporting Israel, one of the NATO ally countries, in the Isreal – Iran war. Hence, for Iran, the USA is one of its biggest enemies and would not like to miss any opportunity to get back at them.

The supreme leader of Iran is an 85-year-old Iranian cleric Mr. Persiani. He has held the post of Supreme Leader of Iran for the last 35 years. He has warned the USA several times against supporting Israel in the recent Israel–Iran war. For him, the President of the USA admitting about a project to kill millions of people is an Allah-sent opportunity. In his long-televised speech, delivered in Persian, he used all the extreme rhetoric possible. He did not hold himself back. He referred to Mickey Cash, Leo Smith and Cyrus Clark as the 'Axis of Evil'. This term has always been used by the USA for Iran and its partner countries. Excerpt from his speech translated into English:

"USA has a history of massacre. The USA is the country that dropped the atomic bomb and killed millions of people in Japan. Now the USA wants to use its newly developed weapon of mass destruction to kill millions and billions of people across the world. We will not let them kill our citizens and fellow Muslim brothers across the world; we will not let them get away with this massacre. Muslim brothers across the world will unite and fight against the USA. We will make the USA pay the price for this. We will also be joining hands with Russia and China to nuke the USA into oblivion."

Iraq

Iraq is a country of around 46 million people, mostly Muslims. The average life expectancy in Iraq is around 72 years. The USA has a history of meddling in the politics of Iraq with its vested interest related to control of oil and gas resources which is abundant in Iraq. Although the USA supported Iraq during the Iran-Iraq war, the USA's support has always been with conditions attached and in its self-interest. The USA has tried to control the politics of Iraq several times by overthrowing popular leaders and establishing its stooges on the throne. The USA was also responsible for the capture and killing of popular Iraqi leader Saddam Hussain under the false pretext of the Weapon of Mass Destruction (WMD). Currently, Iraq has a friendly relationship with the USA. Hence Iraq did not use war rhetoric against the USA. However, Iraq also denounced the USA and warned them against killing any Iraqi national. Iraq refrained from using very harsh words, but they were very clear in their position against the USA's project Auto-off @ 65.

Pakistan

Pakistan has a population of around 250 million people. The average life expectancy in Pakistan is around 67 years, only 2 more than the cut-off year under project Auto-off @ 65.

Pakistan is currently a bankrupt nation seeking aid from all possible countries including the USA, China, UAE and others. The USA has used Pakistan as a military base during the Afghanistan war. The USA also has a history of covertly supporting Pakistan against India as the USA used to consider India to be closer to the USSR. However, things have changed. Now, the USA sees India as one of the key allies in the region, not only because it wants to balance China in the region but also due to trade, technological and other ties with India. The USA never saw Pakistan as an equal and always used Pakistan as a base to balance equations with other countries like Russia, China, and Afghanistan.

Pakistan, considering its current state of bankruptcy and political uncertainty refrained from making any comments. Pakistan has close ties with China and they are begging for aid from each and every major country of the world. Right now, they do not want to upset equations with anybody. Hence Pakistan refrained from comment, as if nothing had happened.

Bangladesh

Bangladesh has a population of around 172 million people and average life expectancy is around 74 years. Bangladesh is currently under political turmoil. Recently, revolutionists led by student leaders have overthrown the Hasina government. The protests were led by students and supported by the general public. It was called the uprising of Students and People. Both groups shared similar objectives related to government accountability and systemic change. Early this year, in July 2024, the unrest intensified, became violent, and turned into a civil disobedience movement, ultimately leading to Hasina's resignation and eventual escape to India. The revolutionists who have assumed power in Bangladesh are trying to pressure India to send back Hasina so that they can punish her. As common in Bangladesh politics, they want to eliminate Hasina so that there is no chance of her returning to power at some later date and taking revenge on current leaders.

Due to the current uprising, Bangladesh's equations with other countries have completely changed. India was a friendly country before the uprising. Then PM, Hasina shared a good relationship with the Prime Minister of India. Maybe, due to relationships with India and the PM, she managed to find exile in India. Bangladesh's new leadership is in the process of spoiling relations with India. They are supported by Muslim fundamentalists and have tacit support of Pakistan. There have been many reports of crimes against minorities, especially Hindus, in Bangladesh which has been condemned by most countries including the USA and India.

Sheikh Hasina's father, the first President of Bangladesh, and most of the family members were killed in 1975 in a coup. Sheikh Hasina somehow managed to escape to India and remained in exile in India for several years before returning to Bangladesh. In a surprising reversal of fortunes, she managed to get elected as Prime Miniter of Bangladesh and served as Prime Minister for more than 20 years, making her the longest-serving female head of state in any country of the world. Hasina's later years were considered to be as autocratic as other Bangladeshi leaders before her. Hasina had recently won the elections in January 2014, but there were doubts about the impartial conduct of the elections.

Due to the uprising, unrest, and coup, Bangladesh's economy has been impacted and they are also in financial trouble, similar to Pakistan. Bangladesh is absorbed in its own troubles and refrained from making any comments against the USA. Like Pakistan, they also want to remain in the good books of all major countries of the world as they also need financial and other assistance.

UAE

UAE is one of the most progressive and wealthy Islamic nations in the world. UAE has a population of around 11 million and the average life expectancy is around 80 years. UAE has a very good relationship with the USA. UAE and the USA have close partners and collaborate on business, technology, and military.

President of UAE, His Highness, was also caught by surprise, like everybody else by the announcement of the Project Auto-off @ 65. UAE conveyed their strong displeasure and disapproval against the project both privately and publicly. They refrained from using provocative, instigating, or insulting language against the USA and its leadership but did not mince words against the USA for taking unilateral action without any consultation with the partners. UAE requested the USA to take back the project and warned of implications if the project is not switched off.

Saudi Arabia

Saudi Arabia is the world's second-largest oil producer and one of the close allies of the USA. Saudi Arabia has a population of around 37 million people and an average life expectancy of around 80 years.

The U.S. is Saudi Arabia's largest trading partner, and Saudi Arabia is the largest U.S. export market in the Middle East. The United States and Saudi Arabia have a history of technical and educational exchange which has benefited both nations, and the longstanding security relationship continues to be important.

Saudi Arabia also conveyed to the USA, privately and publicly their displeasure with the project and requested the USA to switch it off immediately.

Other Muslim countries including Qatar, Indonesia, Kuwait, Bahrain, and others also publicly denounced the project and disassociated themselves from the USA. Most of the countries did not escalate it to the level of war-mongering as Iran did, but most of them disassociated themselves and requested the USA to shut down the project.

NATO countries

The 32 member countries of the North Atlantic Treaty Organization (NATO) are:

- **Europe**: Albania, Belgium, Bulgaria, Croatia, Czech Republic, Denmark, Estonia, Finland, France, Germany, Greece, Hungary, Iceland, Italy, Latvia, Lithuania, Luxembourg, Montenegro, the Netherlands, Norway, Poland, Portugal, Romania, Slovakia, Slovenia, Spain, Sweden, Turkey, and the United Kingdom
- **North America**: Canada and the United States
- NATO has also partnered with countries outside of the alliance, including Bosnia and Herzegovina, Georgia, Iraq, Jordan, Mauritania, the Republic of Moldova, Tunisia, and the United Nations. Israel is also considered a major non-NATO ally.

The combined population of 31 member countries of NATO, excluding the USA is around 625 million people. The average life expectancy in NATO countries is 79 years. 14 years more than the cut-off age under the project. It means on average; the life of each individual would be reduced by roughly 20%. These numbers are not that someone can ignore or disregard. However, NATO countries are facing the problem of greying population more than the rest of the world, maybe with a few exceptions like Japan and South Korea.

NATO was established in the aftermath of World War II. NATO is a collective security system in which all the countries have pledged to fight for each other. NATO pledges that each independent member country agrees to defend each other against attacks by a third party or country. NATO was formed to counterbalance against the Soviet Union but has survived the collapse of Soviet Union. NATO has been in existence for more than 75 years. But now the survival of NATO is at risk.

NATO member countries are peeved with the USA's unilateral decision to go ahead with project Auto-off @ 65. The project means the massacre of millions of their own citizen in the hands of another NATO country - the USA. USA has throughout been the leader of NATO, and around the USA the whole NATO revolves. The USA has in the past, many times, proclaimed unilateral war against other countries, and the NATO countries were forced to join those wars, but those were small wars and the loss of life for NATO members was very limited. Now USA has forced on them a massacre where millions of citizens would be killed without even going to war.

Most of the NATO countries are democracies and none of the member countries can convince their electorate to support this whimsical project of the USA. Countries like the UK have been allies of the USA even before NATO, during the world war. But today none of the NATO countries has the guts, nor intention to stand with the USA. Anyway, Russia has not left them with many options. Russia has singled out NATO member countries to make their stand clear. For the rest of the world, if they do not take a stand, Russia would not mind. However, for NATO countries, Russia has not left that option open. Russia was not ambiguous in mentioning in the televised address by the President of Russia, that any NATO country that does not publicly disown the USA would be considered a partner of the USA and Russia has threatened to nuke them also along with the USA. NATO countries are caught between their voters, Russia and the USA. They understand that their voters are the most important stakeholders in the current situation. Hence, NATO passed a resolution signed by all the members of NATO except the USA. They did not even consider discussing the resolution with the USA before issuing it to the public. The resolution read:

"All the member countries of NATO including Albania, Belgium, Bulgaria, Canada, Croatia, Czech Republic, Denmark, Estonia, Finland, France, Germany, Greece, Hungary, Iceland, Italy, Latvia, Lithuania, Luxembourg, Montenegro, the Netherlands, Norway, Poland, Portugal, Romania, Slovakia, Slovenia, Spain, Sweden, Turkey, and the United Kingdom, excluding only United States of America, hereby announce the dissolution of NATO.

NATO was formed to protect humanity from tyrants, evil forces, and evil states. NATO has successfully fulfilled its mission for the last 75 years. We are very proud of our achievements in NATO and we wish we could have continued longer.

However, the unilateral action of the NATO member country – the USA, against all the member countries of NATO has forced the dissolution of NATO. We the signatories of this resolution cannot stand with the USA and support this switching-off of senior citizens. Yes, we appreciate the problem of greying population, and the problem of low childbirth ratio, but Project Auto-off @ 65 is not the solution. The USA should have discussed this with us before taking such unilateral action. In the last 75 years we have sat together and solved many problems, with joint application of minds, we could have solved this problem also.

Through this resolution, we want to give a clear message to the USA that we no longer stand with you. We, the signatories of this resolution, intend to form a new group, with countries sharing common interests but the USA would not be invited or allowed to join the group.

We also want to give a clear message to Russia, that although we do not stand with the USA, we do not stand with Russia either. We do not believe in war-mongering or threatening to nuke another country and its citizens. We believe the world is on the brink of war and catastrophe. This is a time to sit together and apply our brains to avoid this catastrophe rather than muscle flexing and warmongering. The planned death of our fellow citizens is a crisis, but the crisis is still 5 years away. By brinkmanship and warmongering, we should not prepone the catastrophe.

We, the signatories of the resolution, also resolve to take the matter to the United Nations and demand appropriate action against the USA. We do not believe in war but we also would not let the culprit go scot-free."

Mr. Turin read the press release and laughed his heart out. Finally, he has managed to destroy NATO. Finally, he has managed to isolate the USA. Finally, the USA's back is against the wall, and it is completely isolated. He has achieved what he wanted to achieve through his war-mongering. He never had any intention of going into a nuclear war with the USA. He has won the war with zero collateral damage. But he is not done yet. He wants to destroy the USA. Not with nukes or in war but diplomatically, through sanctions, and via financial, trade, and technological isolation. The USA has been hammering the Russian economy hard with sanctions, and Mr. Turin now wants to turn the tables. He is making notes, in his pocket notepad, of things he would insist on when the matter is taken up in the United Nations.

1. *Death penalty for the conspirators*
2. *Compensation of $1 million to family members of each person who is switched off under the project*
3. *Switching off of Project Auto-off @ 65*
4. *UN control over the project Mars. Russia should have a substantial say*

Chapter 3:
The Fallout

Impeachment by Congress

Mr. Turin is right. The USA is completely isolated. The world is now divided into the following groups:

1. The warmongers – Russia, Iran, China, North Korea
2. The opposers and critics – NATO countries, Australia, Argentina, Brazil, UAE, Saudi Arabia and most of the other countries
3. The mute spectators – Pakistan, Bangladesh, and a few other countries
4. Acknowledgers but not supporters. Requesting for tweak – India

There is not a single country that has raised its hand and supported the USA. The USA is completely isolated. Even the US citizens are not supporting the USA. The US opposition has sensed an opportunity and is gunning for the head of the President. In the past also they have managed to impeach Micky Cash twice. However, they were not able to get the impeachment ratified at the Senate the first time, and the Senate hearing for the second impeachment is due in exactly 30 days. Micky has a majority in the Senate and is expected to sail through in the second impeachment voting also. However, this project Auto-off @ 65 is something absolutely different. He would find it very difficult to find any supporters.

The Congress hearing

The USA Congress called for an urgent meeting at the Capitol. It issued show cause notice to the President of the

USA, Leo Smith, and Cyrus Clark. They were issued notice to join the Congress hearing at the Capitol which would be telecasted live across the world. The US opposition and citizens are also under pressure from warmongers like Russia and other countries to show that they are not standing with their President Mickey Cash.

The urgent hearing of the Congress was called at the Capitol on 23rd January 2025, just two days after the big announcement by Micky Cash. Congress was under pressure to hold the meeting and pronounce the guilty verdict immediately, however, experts begged for 48 hours to prepare for the meeting. They convinced the members that they should not go unprepared, which could allow the three conspirators to wiggle out. A second hearing in the same matter would not be possible if the first meeting bombs.

23rd January 2025

Members of both the House, the Senate, and the House of Representatives are present in the hearing. Not a single member of both houses is absent. Members expect it to be a straightforward guilty pronouncement and no requirement of voting, however, nobody wants to take any chances.

The Congress hearing is being telecasted live in more than 200 countries. There is so much interest in this hearing, that it would go down in history as the most widely telecasted event ever. Even more than any sporting event or any other political event.

The hearing begins. First Cyrus Clark is called for his deposition.

The first question Cyrus is asked is – "What is your exact role in the project Auto-off @ 65?"

Cyrus's body language is very stoic. He is sitting there as if he has been preparing for this for years. Cyrus replies without any remorse, guilt, or second thoughts.

"The idea of the project Auto-off @ 65 was conceived by me in and around 2015. I identified the problem and worked out the solution over the next 5 years. I convinced and got on board both Micky and Leo in 2020 during the Covid crisis."

Congress was not expecting this. Nobody was expecting this. Everybody was expecting a straightforward hearing as the three conspirators had already admitted to their crime, but they were not expecting the admission in the opening sentence itself. After the long pause, Cyrus continued to speak. As if he wanted to give everybody in the room an opportunity to get surprised and wanted the camera to capture the surprise. He appears more like the director of the show, rather than an actor.

He went on to explain why he did it, without waiting for Congress to ask, as if that is the obvious next question. He continued to explain in great depth the problem he has identified – the rising greying population, the falling birth rate, and the deteriorating gene quality of mankind. He explained the whole and supported it with a lot of numbers and data. He had presentations and videos ready; he spent more than 1 hour explaining the whole thing in depth. The members of Congress and the audience watching this live on TV started losing patience. The hearing is without any defined end time, but everybody knows that there is a limit to a human's attention span.

One member from the Congress sitting on the dais, responsible for the deposition and cross-questioning, interjected Cyrus.

"Cyrus, we do not doubt your number or the problem that you have explained. But why such a bloody solution?" He moved his head full 180 degrees to check the reaction to his question, he was expecting an energetic agreement to his use of the word 'bloody' and he was not wrong. He was pleased with himself. People shouted in unionism as if they could not agree more and they also wanted to ask the same question. Cyrus curtly replied,

"I am glad the Congress here has acknowledged and accepted the problem."

People waited to hear what he said next, but he also waited. After a long pause, after ensuring nobody disagreed with his conclusion, he continued,

"Now that Congress has acknowledged and accepted the problem, let me ask you – do you have a better solution?"

There is pin-drop silence in the room. It is a very big hall, there are more than 500 people in the hall, but there is pin-drop silence in the room. Everybody is looking at each other. The ten-member panel sitting on the dais is forced to make an effort to look at the faces of other members. Requesting somebody else to take the lead and offer a better solution.

Noticing nobody is taking the lead, the head of the panel argued – *"if you don't have a better solution, you will kill millions of people?"*

Cyrus retorted – *"So you will let mankind go extinct?"*

Again, there is silence in the room. The congressman was not expecting such an aggressive retort.

He gathered himself back and said – *"we don't agree that it is an extinction-threatening problem, do not try to exaggerate things."*

Cyrus again tried to explain why this was an extinction-threatening problem, why we were already late and could not wait any longer. He again gave an hour-long monologue on the topic. His presentation was very informative and logical, not nobody is here to get convinced. He was again cut short by another Congressman. He asked – *"Tell us what exactly you people have done, and how it can be reversed?"*

Cyrus again gave a short reply – *"It cannot be stopped or reversed."*

Congressman gasped – *"WHAT??!!"*

Cyrus again repeated in a stoic manner – *"It cannot be stopped or reversed. We intentionally programmed it that way, we knew there would be public pressure to ignore the problem and kick the can down the road. But we have run out of the road. Already human genes have weakened substantially, and genes deterioration is not reversable. By gene editing and limiting human age, we have managed to stop it from deteriorating further. You might not appreciate it now, but history would look at it differently. Twenty years down the line, mankind would look at this project as the project that saved mankind from extinction."*

Another Congressman, realizing the first congressman was lost for words, inquired again – *"There is no way to stop it and reverse it?"*

This time Cyrus replied with one word – *"No."*

Congressmen asked again – *"I do not agree, but assuming, you are right it can't be stopped or reversed, can we delay it by another 15 years, instead of 65, can the cut-off can be changed to 80 years?"*

Cyrus again replied curtly – *"No."*

A third congressmen interjected, *"We will form a committee of experts and find a solution to reverse it. We have numerous intelligent people, much more intelligent than them. If they cannot do it that does not mean it can't be done. Let's move on."*

This time Micky lost patience and interjected. *"This is the moot point, there is no moving on from this point. The project has been put into action and it cannot be stopped. We have five years, and we all need to prepare for that.*

We three can lie to you and the world that the project can be stopped and we have stopped it. Then everybody can live blissfully till 1st January 2030. Then, suddenly you will realize that the project was on all the time and it was not stoppable. Cyrus and I would be dead by then, and Leo would have migrated to Mars.

Listen to me carefully – WE NEED TO PREPARE. WORLD NEEDS CYRUS, LEO AND I, TO HELP PREPARE FOR 1st JANUARY 2030.

Leo and the team have to ensure migration to Mars is smooth.

We, the government of the USA have to get the world together and help everybody prepare for this big change. We need to educate them on the merits of the plan and why it is in the long-term interest of Mankind.

We need Cyrus and his team to monitor and ensure technically and biologically everything goes as per the plan.

I know you people are here to crucify us. I know you people would love to give the death penalty to us. We are fine with that. We anyway have given ourselves the death penalty. As we told you earlier also, both Cyrus and I would be the first ones to die. We have 5 years, only 5 years, but the world and the project need us for these five years to ensure everything goes as per the plan.

We ourself propose our sentence:

1. *Both Cyrus and I are to be given the death penalty. However, the death penalty would be in abeyance for 5 years so that we can ensure smooth implementation of the plan.*
2. *Leo to guarantee that every single individual gets a guaranteed free seat on the flight to Mars. Leo also needs to ensure decent living conditions on Mars when people migrate.*

This is all we can offer. Please understand that the world needs us right now. If you impeach me and crucify us today, you might get some joy out of it, but nobody else knows the workings of the project. We three are needed to ensure smooth execution of the project."

Now people are scared, feeling helpless, and have lost patience. There is a lot of cross-talking in the room. Congressmen asked everybody to stay calm and started deliberating among themselves. After a lot of deliberation, questioning, and cross questioning they read out the ruling.

"The committee rules the following:

1. *Micky Cash, Cyrus Clark, and Leo Smith are found guilty of planning and hatching a plan that will lead to a large number of deaths. Hence the Congress recommends:*
    a. *Death sentence for Cyrus Clark for originating and executing the project.*
    b. *Death sentence to Micky Cash for using the office of President of the United States of America to empower the conspirators.*
    c. *Congress recommends to form a committee of experts to find a solution to stop and reverse the project*
    d. *Congress directs Leo to ensure the project Mars is executed in time and everybody gets an assured free seat on the flight to Mars*
    e. *Congress recommends putting in abeyance the death sentence of Cyrus Clark and Micky Cash for 5 years or till a solution is found to stop the project, whichever is earlier.*
    f. *Congress also reserves the right to take any other action against the conspirators as they deem fit, once the expert committee has evaluated the project and its implications."*

Before the meeting Congressmen were hoping and aiming to impeach the President of the USA. The death penalty was not even in their consideration, as they do not have the authority to pronounce the death penalty. That can be done by the courts only. The conspirators themselves agreed to the death penalty. They got more than they were aiming for. But they were not happy. They were frustrated. They were feeling helpless. They realized that they were not in a position to call the shots. They have been blackmailed by the three conspirators. They realized that the conspirators had nothing to lose, and Congress had nothing to gain.

The feeling of frustration and helplessness is shared by everybody across the globe who watched the telecast. Now everybody can see the writing on the wall – THE PROJECT CANNOT BE STOPPED; WE ALL WILL DIE!

The Stock Market

The stock market was volatile ever since the coronation ceremony speech, however, even before the Congress hearing ended the market crashed. Before the Congress hearing, people were hopeful that the announcement of Project Auto-off @ 65 meant nothing and that it would be stopped. However, after the hearing, the optimism vanished. Stock markets across the world were down by around 20%. There was panic in the market. The stock market crash was a clear sign, finally, the reality had started sinking in. Finally, people are accepting what the conspirators have been saying since January 20th. Finally, people are realising that there is no going back. There is no way to stop Project Auto-off @ 65. Finally, they realise - WE ALL WILL DIE!

At first, the stock market crash was across the board. Anything and everything crashed. People were not being selective, they were panicking and selling everything. The stock market crash aggravated the panic and hence market sold off even further.

Over time, people started analysing. Analysts come up with reports with the following themes:

Pharma companies and hospitals will lose a lot of business: This was the most obvious one. If people older than 65 years old did not exist in this world then the pharma companies and hospitals would lose a lot of their customers and hence lot of their business. As people grow older, the probability of illness and disease increases. Analysts did a detailed report on what percentage of the customer base of each pharma and medical company is above 65 years of age and what percentage of revenue of these companies are coming from these people.

Among pharma companies, most of the companies were given a ‘SELL” rating. Pharma companies which specialized in medicine for the very old or their product portfolio had a higher balance of medicines targeted to older people like patients of Alzheimer's were marked “STRONG SELL”.

Among hospitals, except for some specialist chains like childbirth centres, neo-notal centres, and sports injury treatment centres, almost all other hospitals and medical centres were assessed to be significantly worse off and were marked ‘SELL”.

Analysts noted that medical research budget and expenses would substantially fall for medical companies as people will switch off at a healthy age. However, they believe this would lead to even further erosion of the value of medical companies.

Associated industries in medical business: Most of the associated industries in the medical business including medical equipment, nursing service, old age homes, pharmacy chains, medicine delivery service, medical testing centres, medical record-keeping services, and medical software enterprises, all were marked “SELL” by the analysts. Even suppliers of non-medical goods like sanitisers, masks, and gloves were marked “SELL”.

Medical insurance: Generally medical insurance is taken by people of all ages. Young people take medical insurance and seldom end up making claims, while old people make a higher percentage of claims. Hence, generally, young people cross-subsidize the old people. Analysts had differing opinions about what would be the impact of Project Auto-off @ 65. Few analysts believe that young people will continue to buy medical insurance, but the claims will substantially come down as older people would vanish. They were of the opinion that due to the reduced number of claims, the insurance companies would become very profitable. Few analysts went even further and mentioned that the cost of medical procedures and the cost of medicines will substantially reduce in future due to limited demand. Hence many analysts rated medical insurance companies as 'BUY'.

On the other hand, few analysts differed in their opinion. Many young people will stop buying medical insurance, they argued. They believed that if medical incidences in the world will come down so will the fear of medical incidences. Young people would be more confident and would be willing to take risks of making it to 65 without any major medical expenses. Hence this set of analysts maintained a "SELL" rating on medical insurance companies too.

Life insurance companies: Life insurance companies were quick to react. They stopped issuing life insurance with insurance coverage beyond 64 years of age. They initiated the process of revising old life insurance agreements too. Life insurance companies were staring at a very high number of claims on 1st January 2030. Many of these companies were anyway not offering any life insurance beyond 60 or 65 years of age. They were better off; however, few have offered cover up to 70 years or even beyond. They are looking for loopholes as the current situation is entirely unimaginable and they could not have anticipated this. The insurance risk calculation has not factored in such an eventuality. They are hoping that the government will give them relief and let them claim deaths due to Project Auto-off @ 65 as an exception to life insurance. Few life insurance documents have exceptions for death due to terrorism. Insurance companies want the project Auto-off @ 65 classified as an event of terrorist activity or terrorism. They are highlighting the speech of Mr. Turin, Mr. Lian and many other leaders across the world where they have used the term 'massacre' for this project. In the near-term Life insurance companies are marked as "SELL" as a lot of claims and litigation is expected. However, the same analysts believe that in the long run, Life Insurance companies will gain due to a falling number of claims.

Consumer goods companies specifically targeting senior citizens: Adult diapers, walking sticks, hearing aids, spectacles, all companies where the clientele was primarily senior citizens above the age of 65 have been marked as "SELL" by the analyst.

Travel & hospitality companies: Surprisingly travel industry was marked as one of the biggest beneficiaries of the recent development. Some analysts argue that 65 plus formed around 10% of the population on the earth. The percentage is even higher among the developed countries where average life expectancy is higher. They argued that a 10% fall in population would reduce capacity utilization in the industry and lead to a fall in business. However, a big group of analysts reasoned that people will travel more. They believed, that with clear visibility of the life end date, people will live it up. People will travel to complete their bucket list. People will travel to enjoy the last days of their lives; people will travel to meet friends and family. Few analysts went further and forecasted that people will retire early and spend on tour and travel so that they can live up the last days of their life. They even added that saving in medical expenses of a family would be added to the discretionary spending budget and would be used for travel and holiday. Most analysts ended up giving a "BUY" rating to travel and hospitality companies including airlines, hotels, restaurants, entertainment parks and recreational centres. After the analyst reports, stock prices of airlines, hotels and other hospitality companies improved. They are not yet at levels before the coronation, but it has substantially improved and recouped most of the losses.

Space technologies companies: Analysts predicted a mad rush for space technologies companies. They predicted that space travel, space transportation, space research and all kinds of industries related to space would attract a lot of investor money. They will also attract a lot of business as 10% of the population of the world would be forced to fly out to space to survive.

For Leo's aerospace company – SpaceY, analysts have differing opinions. Few believe that SpaceY would not be able to make money as they have promised free one-way tickets to Mars to 10% of the population of the Earth. Others believed that SpaceY would still make a lot of money due to the scale of its business. Analysts believe that SpaceY will continue to be the leader in the industry and will mint money due to first-mover advantage, due to scale, and due to government support. Leo's rival billionaire, Ted, founder of the biggest online retail chain in the world, with whom Leo at times competes for the 'richest person on the earth' tag, is also competing with Leo in the race to space. Their rivalry is so fierce that it has become impossible to contain it within the Earth's atmosphere. It has escalated to much beyond Earth, to space. Ted is also building his own space travel company called Green Origin. Currently, Ted is far behind in the space race, however, analysts have given a "STRONG BUY" rating to both SpaceY and Green Origin and several other space startups.

Impulse purchase consumer goods: Analysts believe that although a 10% fall in the world population will lead to a substantial fall in demand, but impulse purchase consumer goods industry including chocolates, ready-to-wear garments, etc will continue to do good. They believe that with savings in medical expenses of households, people would spend more. They believe there would be an overall fall in saving rate in the world as people will not have to plan for a very long post-retirement period. Analysts believe that a definite death date will encourage people to spend more. They have given a "BUY" rating for these companies.

Baby care and toys: There is an agreement among analysts that the project Auto-off @ 65 will encourage people to have more kids and have kids early. Analysts are of the opinion that the birth rate in the world has been falling due to the increasing burden of supporting the old generation on the shoulders of earning members of the family.

Analysts also forecast that people will stop delaying having kids as they know that their life span is limited to 65 years. They would like to have kids between the ages of 25 and 35 so that, kids have completed their studies, become independent and married before they get switched off from earth. Hence analysts gave a "STRONG BUY" rating to companies in child care, baby care, toys and similar industries which has infants and their parents as their primary customers.

Condoms: Continuing with the theme of "living up the life," analysts believed that the use of condoms would increase as people have more sex. They also believed that the vanishing of senior citizens aged above 65 would have no impact on the condom industry as they anyway don't use condoms. Some people argue that if we are forecasting that people will have more kids, then we cannot forecast that people will use more condoms. That would be contradictory. However, such arguments were discarded as shallow and the common consensus was that it would lead to better prospects for the condom industry. Hence, they gave a "BUY" rating to companies dealing in condoms and similar products.

Luxury goods: Continuing with the theme of 'living up the life,' analysts believe that people will save less and spend more. Analysts believe that people specifically in the age bracket of 55 and 65 will 'live up' their lives. They would not care to save much as they would be aware that they would be switched off at 65. Analysts gave a "BUY" rating to these companies.

Vice products – Liquor, cigarettes, drugs: Analysts went all in with their 'live it up' or 'living up the life' theme. Some even named it the '*Zindagi Na Milegi Dobara*' theme. There was a consensus among analysts that vice products would do well. They forecasted that people would be less scared of old age; they would abuse their bodies more. They will use the vice products without restraint as in either case they can't live beyond 65. Vice products also got a consensus "BUY" rating.

Government spent: Analysts predicted that with the vanishing of people aged above 65 years from the face of the earth, governments with end up saving a lot of money. Today a big part of the government budget goes into medical and retiral benefits spent on senior citizens. This is specifically true for certain European and other developed countries which provide free medical facilities to their citizen. Saving in government budget on medical and retiral expenses is expected to make the government budgets much healthier. For the last couple of decades government budgets and finances and been deteriorating across the world and more so in developed countries. Project Auto-off @ 65 is expected to be a major reprieve and boost for the government budget all across the world. It is expected that the government will end up spending the savings on medical and retirals on infrastructure and education. Hence, it is expected that infrastructure and education sector companies will do well. However, analysts avoided giving a clear "BUY" call on the infrastructure sector and education sector. Across the world, governments cannot be trusted with surplus money. They might end up wasting it on defence or doles. Some are of the view that better government finances will lead to lower taxation and overall industry and economy will benefit. Others think that is wishful thinking, they believe government will find ways to waste money and keep government budgets under strain.

Most analysts and market experts forecasted doom for the medical industry and a boom for the 'live up the life' industry. However, there are contra forecasters also. Few experts disagreed that people will be able to 'live it up' in the short run, in the first 5 years leading up to doomsday of 1st January 2030. They believed that humans have not gotten used to Auto-off @ 65 and will not get used to it anytime soon. They argued that in the next 5 years, people will suffer from a lot of anxiety, and mental issues. People will behave erratically, commit suicide, harm themselves, and harm others. These experts believe that waiting for death is not easy. Certainty of death will impact different people differently. In the early years, it will lead to a lot of anxiety and mental disease. It will lead to a lot of law-and-order problems. In the long run, mankind might get used to the certainty of death and respond positively to that. Experts believe that will take a minimum of 10 years.

Expert Committee of WHO

Leaders from across the world had many bilateral talks and group calls. There is panic all across the world and leaders need to do something. At least they need to show their citizens that they are doing something. After the US Congress hearing where the three conspirators curtly repeated again and again that the project Auto-off @ 65 cannot be reversed or stopped, there is a feeling of helplessness among the world leaders. However, they cannot show that they are helpless. Hence, they continue making speeches and noises against the project. They kept cursing the three conspirators and the United States of America.

The leaders across the world had to do something concrete to show that they were doing something. Also, they still harboured some hope that something could be done. Maybe some bright minds and some scientists can come up with some solution. Leaders of several countries decided to form a committee of bright minds to study and find solutions. They also resolved to have close coordination of expert committees across the world so that there is knowledge sharing.

Most countries created an expert committee of 5 to 11 members to study the damage done by the three conspirators. They were tasked to study exactly what Cyrus Clark and Leo Smith and their team have done to edit the human genes and find a way to reverse it or re-edit it. The brief was very simple: Find a way to stop or reverse the Project Auto-off @ 65.

For close coordination between these expert committees in each country, an apex committee of experts was created under the direction and guidance of the World Health Organization (WHO). WHO is a specialized agency of the United Nations (UN) responsible for global public health. It is headquartered in Geneva, Switzerland, and has six regional offices and 150 field offices worldwide. Most countries are members of the WHO, and it is the largest inter-governmental health organization at the international level in the world.

The apex committee is comprised of 7 experts from across the world. The committee was provided with an office in the headquarters of WHO in Geneva. The apex committee was tasked with coordinating with all expert committees across the world and coming up with a solution to neutralize Project Auto-off @ 65. The committee was promised unlimited resources and unlimited access to top leaders of the world.

As soon as the apex committee was formed, they got into action. They did not lose much time as time was of the essence. Although there are still 5 years left to doomsday there is panic all across the world. Citizens want their leaders to do something, they want to see their leaders doing something. Also, gene editing and biotech is a difficult and slow science. Nobody knows how much time it will take to come up with a solution, test the solution for any side effects, first on animals, then on humans, and then go into production. Distribution of the solution is another challenge. The world has recently experienced all these challenges during the COVID crisis. The challenge this time is equally time-critical and has equal global impact if not more. Many of the experts who worked on the COVID crisis have been involved in this project so that their experience of global-level crisis management can be leveraged.

The first thing the apex committee requested was access to Cyrus Clark along with the other conspirators. They also asked was all the files, research, and documentation relating to the project Auto-off @ 65. Representatives of WHO immediately communicated the request for information to the President of the USA, Cyrus Clark, and Leo Smith. Surprisingly, the three conspirators were unbelievingly forthcoming in sharing information about the project. They have everything ready on the cloud and as soon as the request was received, they provided access to the data to the apex committee.

The apex committee was not expecting this. They were expecting resistance to the sharing of information. They were expecting the sharing of limited information, they were expecting a lot of back and forth. The instant sharing of information by the conspirators was a pleasant surprise for them. The apex committee now has the task in hand to decide how much of this information needs to be shared with the expert committees of various countries.

The apex committee started going through the data shared with two purposes in mind. One is to get a high-level understanding of the issue on hand so that they can prepare a plan of action. Second, to decide which part of the data can be shared with the respective country's expert committee. Each country had its own reservations about sharing this information with other countries. People in power at WHO were worried about sharing this information about gene editing and satellite functioning with so-called 'rouge' nations like North Korea, Iran, Russia, and China. The apex committee again reached out to Cyrus Clark to pick his mind on what information should be shared and what should be kept classified.

The deliberation on what should be shared and what should not be shared ended up becoming the main job of the committee. There was a lot of to-and-fro. Heads of states of various countries got involved. Expert committees of these countries were elucidating to them that this information is very critical for them and is the starting point. Without the complete information, they would be working in the dark and their chances of success would be negligible. Limited information sharing became a topic of hot debate between countries and at the UN. Due to a lack of trust between countries, the critical project was suffering, the mankind was at the risk of losing the bus. In a very short duration of less than 5 years, the world is facing the same issues again. First with Covid and now with Project Auto-off @ 65. Everybody is supposed to fight for mankind, but humans cannot trust each other and hence fight among themselves for access to information. More than a month was wasted in this infighting.

Cyrus Clark and the rest of the conspirators got the sense of the fight. They were frustrated that human never fails to show their standard human behaviour. The behaviour of fighting among themselves even in the worst of times when working together is very critical. Cyrus went ahead and gave access to all the files to all the expert committees across the world. Effectively they made the data public, eradicating for once and for all the cause of the fight between the nations. Cyrus anticipated that next the countries would fight over direct access to him and his team. These experts would like to question, interrogate, interview, and cross-examine Cyrus and his team.

Cyrus himself had to focus on the successful implementation of Project Auto-off @ 65. He was not in a position to answer the same questions from different expert committees across the world. Hence, he offered to make himself available at the Geneva headquarters of WHO to answer all the questions. He offered to be there for one full week between May 1st and May 8th of 2025. He offered to answer all questions on behalf of the three conspirators. He also assured that the other two conspirators would be available over video conference as and when required, however, he assured that he is fully capable of answering all the questions by himself and would not require the help of the other conspirators.

He attached two conditions to his offer. One, the meeting should be open to all expert committees from across the world. During these seven days, all expert committees should be given a chance to ask their questions and satisfy themselves. Cyrus does not want to waste his time answering repeated questions one by one to different committees. Second, he insisted that the questioning should be live telecasted and recording should be made public. He wanted to ensure 100% transparency. Members of the apex committee were again taken by surprise, they did not expect Cyrus, who has the reputation of keeping all his technologies, patents, and research close to his chest to insist on transparency. Even during the Covid crisis, when the time was critical and millions of lives were at risk, Cyrus was against making the Covid vaccination formula patent-free. Everybody was surprised by the sudden change in the nature of Cyrus.

The WHO and the apex committee agreed to Cyrus's demands and the meeting was fixed for 1st May as suggested by Cyrus. The apex committee and the various expert committees across the world emersed themselves in the mountain of information shared by Cyrus. They wanted to be prepared for the meeting. They knew they did not have much time. They knew they would not get access to Cyrus again and again; they had to make the most of it.

Even before the meeting on 1st May 2025, there were rumours of many expert committees across the world confirming their biggest fear. Cyrus is right, Mickey is right. The project cannot be stopped. The project cannot be reversed. However, none of the expert committees, WHO, or the country leadership confirmed these rumours.

On the first of May 2025, Cyrus was present at the Geneva headquarters of WHO sharply at 8.30 am. The meeting is supposed to start at 9 am, but experts from various countries across the world have been flying in for the last 2 days. There are some experts who are flying in, in the morning. Cyrus used his charter plane to fly from the USA and reached well in time.

The meeting started at 9 am. The first 30 minutes were spent in the introduction of the members of the apex committee and a few other renowned experts in the room. A few ministers and high-ranking government officials from various countries were also present in the room. Expert and bureaucrats had their army of support team with them. Some were allowed entry into the room others were watching in the other hall over television. The event was being live telecasted across the world and experts from across the world have also joined in through video conference.

Cyrus, since his days as head of MacroSoftware, has a reputation for having a battalion of support team for important meetings. The apex committee was expecting Cyrus to come to this meeting with at least 20 team members. To their surprise, he had only 3 people there with him.

After the introductions and brief chat, the chairman of the apex committee started off the meeting by stating the agenda of the meeting, rules of conduct, and what they aim to achieve in the next 7 days.

"The apex committee has been formed by WHO with the single objective to stop or reverse the Project Auto-off @ 65. The members of the apex committee of WHO, experts of various expert committees from across the world, and government dignitaries from various countries have gathered here to interview the creators of the project. The creators of the project are represented by Cyrus Clark, the founder of MacroSoftware and Clark Foundation. The other two creators of the project, Micky Cash, the President of the United States of America, and Leo Smith, founder of SpaceY, SatelliteLink, and Nikla, are not present in the meeting but are available to join via video conference when required.

Mr. Clark has assured us that it would not be necessary as he has all the required information of the project and he is willing to transparently share it with us all. Mr. Clark has already shared the detailed project information with us, in advance of the meeting and we would like to put on record our acknowledgment and appreciation for being forthcoming. We expect him to be equally forthcoming over the next 7 days when members will be asking various questions and expect detailed explanations.

The event is being telecasted live across the world. Hence, we would request each and every member to maintain decorum, use dignified language and tone, and give proper respect to everybody including the so-called 'conspirators' of the project. Setting an example of dignified language, we will use the term 'Creators of the project' to refer to the above-mentioned gentlemen and we request everybody to use the same term. Anybody not following the discipline would be made to leave the room.

For the audience watching the deliberation on television, we would like to inform you in advance that the discussion is expected to be very technical and detailed. We expect that there will be use of a lot of jargon and technical terms. Most of the audience might not find it interesting or easy to understand. We will make no effort to simplify the discussions to make them more understandable for the audience. However, we will endeavour to release a minute of the meeting at the end of each day where highlights of the meeting would be shared in simple terms for the consumption and understanding of the common public."

The chairman then fired the first question to Cyrus. *"Let me directly come to the most important question - you and the other creators of the project have been repeatedly emphasizing that the project cannot be reversed or stopped. Explain to us why it cannot be reversed or stopped."*

Cyrus politely replied – *"Because we created it this way. One of the most important objectives of the project was that it should not be reversible or stoppable. The whole purpose of the project would have been lost if it was stoppable. I hope the committee will approve when I say that the creators of this project are very smart people. We knew that when we announced the project there would be a lot of resistance. We knew that if there was even an iota of possibility of this project being stopped, experts from all over the world would get together to stop it. Hence, we ensured that there was no possibility of stopping the project. The project goals are very important for the survival of mankind, and it was our responsibility to ensure that this project is executed, without any stoppage. Completed, even if resisted by the same mankind for whom this project is designed."*

The chairman fired his second question – *"I am diverting a bit, but one thing has been troubling me for a while, so I will ask that first - If the creators of the project were so hell-bent on making the project unstoppable and irrevocable, why did you people disclose the project. You people could have just waited for 1st January 2030 when the project effects would have been auto-triggered. Why risk experts from across the world getting together to stop the project? Why go through the loss of image and reputation that the creators have gone through in the last few days? Why risk being crucified for what you have done when you could have done it even without anybody knowing it?"*

Cyrus smiled and replied – *"Finally, this is the most intelligent question I have been asked in the last three months!"*

Cyrus took a few seconds to observe the reaction of people around him, especially the non-technical politicians and bureaucrats in the room. He took his own sweet time and then gave a very short reply. He purposely kept it short to increase emphasis, to ensure that the message was not lost, and to ensure that everybody heard it loud and clear.

"BECAUSE WE WANTED TO GIVE PEOPLE TIME TO PREPARE."

He again took a long pause. Waited for everybody to digest what he just said. It should have been obvious, they mentioned it in the first speech itself, and they have repeated it several times. He hoped that this time people will get it. After the pause, he continued:

"We wanted people to prepare. We don't want people to die suddenly leaving their loved ones in a lot of sorrow and uncertainty. We wanted people to get enough time to get their things in order before switching off from Earth.

Death is a certainty. Everybody will die. We have not made it that way, God has made it this way. However, the timing of death is an uncertainty for mankind. This uncertainty of timing gives the false illusion to mankind that it is possible to live forever. We wanted to give certainty to timing also. Currently, death's timing is unpredictable and hence it is feared. We want to take away the fear from death. We wanted to give certainty to the timing of death. We want mankind to embrace death as any other fact of life.

We intentionally did not switch off everybody who was above 65 years old on day one. We gave everybody 5 years to prepare. Let me emphasize again so that the message reaches everybody:

PREPARE, PREPARE, PREPARE, and then EMBRACE DEATH."

The chairman of the apex committee was not convinced with the answer but he understood the selfless motive behind making the announcement and appreciated it. However, he did not want to communicate his appreciation in words as that would not be taken well. Most people in the room want to crucify the creators, even a small appreciation by him might make him appear biased towards the defendants.

Chairman fired the next question – *"It is good that people are getting an opportunity to prepare, but why do people have to die early, why can't people die their natural death? Just to provide an opportunity to prepare and add certainty, why reduce their life?"*

Cyrus appeared a bit frustrated with the question, but he was quick to control his frustration. He realized that the buck stops at the doorsteps of the creators. He understood that the creators themselves are at fault that they have not been able to explain this moot point yet and people have to ask it again. He made a mental note to himself that he needed to answer this question in detail and answer it so well that nobody had to ask this question again. He started explaining:

"Anybody who is born has to die. Death is a certainty; God has made it this way. Even God, belonging to any religion, who came to Earth as an Avatar or messenger or in any other form, died and left the Earth. Death is the cleaning process that enables the birth of the next generation.

In agriculture, the plants are there for a fixed life, generally a season, a year, or a few years. We plant the seed, let it grow, harvest it and after a while uproot it to sow new seeds. We do not leave the old plant in the ground. We do not let it die or rot. Just before the sowing season, we uproot the old plant to sow new seeds. Killing of plants is also killing living beings. Still, we kill plants. Because if we show remorse for the old plant and try to do the next sowing without uprooting the old plants the new plants will not be able to grow properly. Most of the new seeds we plant will have to compete for the resources with the old rotting plant. In the case of plants, the critical resources are water, air, sunlight, and minerals from the Earth. For humans also the critical resources are water, air, sunlight, and earth. A new seed will not be able to compete with an old plant in the race for resources as the old plant is already well entrenched in the earth, while the seed is too feeble at the start. Seed will lose to plant. The upcoming generation will always lose to the old generation in the fight for resources, as the old generation is already in control. Within three or four harvest cycles the quality of harvest or the plant will deteriorate beyond repair. The quality of the seeds will also deteriorate beyond repair. The plant would go extinct. Similarly, MANKIND WILL GO EXTINCT, IF WE DO NOT UPROOT THE OLD PLANT IN TIME."

He waited for a few seconds and again emphasized: "*THE MANKIND WILL GO EXTINCT IF WE DO NOT UPROOT THE OLD PEOPLE IN TIME.*"

The Chairman interjected – "*You keep repeating that Mankind will go extinct, but mankind has been there for thousands of years, it has prospered for thousands of years, what has changed now? Why suddenly do we need to switch off people at 65?*"

Cyrus replied: "*Average life expectancy of humans was not more than 65 years until a couple of decades ago. Only in the last couple of decades, due to the advancement of medical science, the average life expectancy of humans has crossed the dangerous level of 65. If it was not for the artificial intervention of medical science, nature was ensuring the cleaning process."*

The Chairman and the apex committee have now started to see the point. Although they are not convinced, not converted, but started to see the point. The Chairman wanted to understand the plan better. He continued with his next question – *"I would not equate human life with plant life but you have made a fair point. But why do humans need to die at the age of 65? What is the logic behind the number 65? Why not 75 or 85? What is the scientific basis behind 65?"*

Cyrus again appreciated the question – *"I am glad that you asked this question. This is also a very important question that needs to be answered.*

Tell me – why retirement age across the world is around 60 years? Few years here and there, but in most countries the retirement age is around 60 years. What is the reason for that?"

Cyrus paused. He waited as if he was expecting the chairman, or the members of the committee, or somebody in the room to answer that. Nobody in the room risked doing that, and Cyrus was also not expecting anybody to really answer that. He continued: *"Let me answer that. Because 60 years is the proper productive life for a human being. In these 60 years, the human being grows, educates, marries, and reproduces. The human leads a productive life and then starts his or her march towards death.*

A female member of mankind, whom God has entrusted the most important responsibility of giving birth and ensuring the continuation of mankind achieves menopause after 50 years or so. With all the years of research and development, we have managed to substantially extend the average life of a human being but we have not been able to extend the fertility period. There is enough data there to show the correct age to bear child for a female is between 25 years and 35 years. When people delay childbirth beyond 35 years, the chances of having a healthy child decrease. The chances of the child being born with diseases increase. If we continue to delay childbirth, the seed will continue to deteriorate.

60 is the retirement age because that is the end of the productive age of human beings. There may be a small percentage of people who might remain healthy and productive even after 60, but when you are sowing new seeds, you have to get rid of all old plants. We have given an extra 5 years so that people can enjoy the fruit of their productive age. So that they can enjoy 5 years of retirement.

Post 65 years, they are a burden on the earth, they are a burden on the next generation, they are a burden on themselves. Post the age of 65, life is not life, it is suffering. Medical science has extended the life of human beings but the quality of life after 65 is not great. By extending average life beyond 65, medical science has not extended life, it has extended suffering. We all need to ask ourselves if it is worth living after 65. Today many old people continue to live a diseased life, full of suffering, because they do not have any other option. They don't know any other option. We have not given them any other option. The plant cannot uproot itself; it needs to be uprooted.

As Mr. Mehta, the Prime Minister of India, mentioned in his speech, even as per ancient Hindu texts, the learned have suggested that after a certain age, human beings should take up 'Sannyasa'. The texts have also hinted the age to be around 60. Human beings should move out of the world where they will compete for resources with the next generation. Moving to the jungle was an option in the ancient texts, where you embrace death as you know what is the probability of a sixty-year-old surviving in a jungle. We, the creators of this project, have created a new but similar option. All thanks to Leo, mankind now has the option to leave the earth and the resources of the earth to their next generation and move to Mars."

The Chairman of the apex committee was impressed with the answer. He mentioned *"Mr. Cyrus, I would love to read through the Hindu texts you mentioned to understand the concept of Sannyasa and the philosophy behind it. But I have to grant it that there is merit in your argument. Regarding the option of migrating to Mars, I find it very interesting. I would love to understand that better, but that is beyond the scope of our mandate. Our mandate is limited to finding ways to stop the death of people. Different bodies of the United Nations might discuss the option of Mars with you.*

I now call up other members of the committee, people in the room, and people joining via VC to ask their questions."

Other members of the committee and others present in the meeting asked several technical and non-technical questions directed at Cyrus. Cyrus explained with a lot of patience and in-depth. He was making a very conscious effort to simplify the answers so that the common man watching the live telecast on television could also understand and appreciate the vision of the creators of the project.

Questions included:

1. *Explain how you managed to edit the human genes so precisely that people will die exactly at the age of 65?*
2. *How you have managed to ensure that future generations, who are not born yet would be born with modified genes?*
3. *Explain the role of SatelliteLink in the whole project.*
4. *Is there a possibility to make exceptions for a few select people?*
5. *What is there specific to Earth's environment that people will die here but would survive on Mars or other planets?*
6. *Explain why it cannot be reversed. Which genes you have tampered with? What changes you have made? Why we cannot again tamper with the same genes to increase the switch-off age to say 100?*

People kept asking questions and Cyrus kept replying to the questions. The interview went on for 7 days as planned. Cyrus replied each and every question, with patience even if they were repetitive. His replies were very detailed, technical, and precise. Some people agreed on the requirements of the project some didn't. Some people agreed with the execution philosophy some did not. But everybody agreed on one thing – Cyrus has replied to all their questions with full sincerity and honesty. He has been very detailed, and precise and backed everything with facts and numbers. He has been very transparent and forthcoming.

At the end of the day, the apex committee appointed by the WHO circulated the final minutes of the meeting for all 7 days along with its comments and recommendations. It read as follows:

1. *"Mr. Cyrus explained in detail why the project cannot be reversed or stopped. Most experts agree that the answers appear convincing and technically sound. However, we will continue to find a solution.*
2. *Mr. Cyrus convincingly explained to the committee that the new generation can grow and prosper only if the old generation does not compete with them for resources.*
3. *Mr. Cyrus convinced the committee that the age of 65 is the correct age for the old generation to make way for the new generation*
4. *After 7 days of deliberation and discussion, the committee was able to appreciate the genuine motive of the creators of the project and would recommend to the United Nations, to debate and discuss: WHETHER WE REALLY WANT TO STOP THE PROJECT?*
5. *Finally, through these minutes, the committee would like to give prominence to the point on which Mr. Cyrus harped again and again. Mankind needs to prepare for the new world triggered by this project. We have only 5 years; we should not waste time."*

United Nations – Divided Nations

Even before the WHO meeting, it had sunk in across the world that Project Auto-off @ 65 is unstoppable and it is unreversible. It has sunk in that the 1st of January 2030 is the doom day when everybody above the age of 65 will be switched off from the Earth. Either they will fly off to Mars or they will die. THERE IS NO OTHER OPTION.

The Committee of scientists and experts who were appointed to find a solution to stop project Auto-off @ 65 have also confirmed that it cannot be reversed. The gene editing cannot be reversed. People are surprised that the creators of the project have managed to convince the expert committee, compromising of the best minds in the world, that there is an urgent need to ensure the next generation does not have to compete with the older generation for resources.

Post the recommendations of WHO, the United Nations (UN) called a meeting of all the member countries for discussion. As the project impacted each and every country in the world, it also gave opportunities to non-member countries to send an observer to the meeting,

The meeting was scheduled on 15th July 2025 at UN headquarters in New York City, USA. The stated agenda of the meeting was:

1. *Deliberate and discuss the demands of certain member countries including Russia and China. The demands as notified in writing to the UN include:*

   a. *Castigation and punishment for creators of the project*
   b. *Penalty and sanctions for USA*
   c. *Compensation to family members of the people who will lose life under the project*
   d. *UN control over the project Mars. Russia should have a substantial say*
2. *Discuss and deliberate on the WHO Apex Committee recommendation – especially the recommendation to discuss whether efforts should be made to stop/reverse/modify the project Auto-off @ 65 or if it should be allowed to run as it is.*
3. *If the project is allowed to be run as it is, what kind of preparation is required, and how the world will go about it?*
4. *Considering the sudden urgency and increased importance of the Mars colonization project due to the project Auto-off @ 65, what member countries can do to help expedite the timelines of the project?*

UN was planning to telecast the meeting live on television across the world considering the impact these discussions will have on the life of each and every person on earth. There was a high level of anxiety among the people and hence people wanted to witness the discussion. However, some of the member countries, especially the powerful member countries, the ones who have made demands and hold veto rights, vetoed live telecasts.

The meeting started at the scheduled time with all major countries including the USA, China, Russia, India, UK, Germany, Australia, Canada, and France participating. Most of the countries were represented directly by the head of the state. Mr. Mickey Cash, President of the USA, Mr. Jia Lian, President of the Republic of China, Mr. Turin, President of the Republic of Russia, Mr. Mehta, Prime Minister of India, and many other world leaders were present.

As the meeting was called on the initiation of Russia, Russia was given the chance to make the first speech. Mr. Turin made a long speech cursing the USA and demanding humongous penalties and sanctions against the USA. Excerpts from his speech:

"The USA has pushed mankind to its biggest crisis ever. More than a billion people would be killed by the USA in the next 5 years, thanks to their foolish project Auto-Off @ 65. Russia demands the following:

1. *Financial, trade, and technological sanctions against the USA.*
2. *Public crucifixion of the three creators of the project*
3. *Compensation of $ 1 Million to relatives of people who are massacred due to this ill-conceived project*
4. *Military sanctions against the USA. USA should be forced to withdraw their troops from across the world and close down all military bases outside the USA.*
5. *Social, financial, trade, technological, and military isolation of the USA.*
6. *UN control over the project Mars. Russia should have a substantial say."*

China, Iran, and many other countries agreed with the Chinese demand and repeated the demands. Few other not-so-powerful countries only demanded compensation and conveniently omitted the rest.

After speeches by countries making the demands, the UN allowed Mr. Mickey Cash, President of the USA to speak on behalf of the USA and the three creators of the project. For the last few months, the world has seen a very docile version of Mickey Cash. Generally, Mickey Cash is known to be the most brash President of the USA ever. Mickey is known to not follow protocols and not follow the stated stand of the USA. He is so unpredictable that he can suddenly change his stance on any matter. He is known to treat the presidency of the USA like a business role. He frequently gets into negotiations with other countries without caring a damn about the moral stand of the nation. He is known to believe in value more than values.

Micky is 6 feet 3 inches tall. He is well-built, like a boxer. He also has the attitude of a boxer. He is not known to hold back his punches. He has funny yellow colour hair and a slight paunch more like a sumo fighter than a boxer. He can be a bully, he can be a negotiator, but he always keeps the upper hand. He has a very limited vocabulary, but his limited vocabulary does not limit him from expressing his contempt for his rival. The speech of Mr. Turin, President of Russia has completely annoyed him, and he is all set to give it back. He starts his speech.

"America is a great country. Greatest country in the world. America has protected mankind for hundreds of years and will continue to protect mankind. Protection of mankind demands sacrifices and America has never shied away from sacrifice.

This time protection of mankind is demanding sacrifices from senior citizens of the world. As a true American, I am willing to make the sacrifice. I am willing to die to save the world.

We demand the same from others. I trust Mehta, he will not shy away from sacrifice. I demand the same from Turin and Lian.

The Russians do not have a track record of being very bright. Over the last few months, we have tried to explain to the world why this project is necessary to save the world. Even the Apex Expert Committee of WHO, consisting of the brightest minds in the world, is now convinced. We cannot do much if the dimwits do not understand.

Specifically taking about the demands – we do not agree to any of them. They are rubbish! Total and utter rubbish! We on the contrary demand from Russia and China to pay their share of the cost to make the project successful. We need to spend a huge amount of money, huuuuuuge amount of money to colonize Mars. As we are providing Russians and Chinese citizens free transportation to Mars, we demand Russia and China to bear the cost of their citizen's transportation and stay.

Experts have now agreed that this project is unstoppable. Hence it is in your interest to ensure that the project gets full support and we are able to colonize Mars as soon as possible. Any country that tries to impede the process, does not share resources, does not share technology, etc will be banned from the Mars project. Citizens of those countries would not be allowed to board the Mars spaceship. It is now the responsibility of citizens of the respective countries to put pressure on their governments to support the USA. If you are not able to push your government to support the project, you will miss the flight to Mars.

America has been protecting NATO countries for more than 75 years now. Leaders of the NATO countries chickened out at first sight of trouble. America has made many sacrifices for you, and you chickened out when mankind needed you to make some small sacrifice. I don't care. Now Europe and Russia are your problem, not mine. Best wishes.

Let me make it very clear, any country that does not cooperate with America, does not support America, or tries to start a conflict with America will pay a heavy price. Citizens of those countries will not be allowed to travel to Mars. Now it is the responsibility of the citizens of respective countries to get their leaders in line or vote them out. Most of these rouge nations do not even have democracy, you will have to throw out your autocratic leaders.

If you call it blackmail, let be it. Yes, I am blackmailing the world to embrace the project Auto-off @ 65. You do not have any other option. If you are bright enough you will appreciate that it is in your interest. If you are not bright, consider it blackmail. Either way, you do not have an option but to embrace the project. Focus your energy to plan for the new world, America promises you a much much better world."

Everybody in the hall is shocked. Suddenly they realise America still holds all the trump cards and they are helpless. America can blackmail them; America can bully them. America can dictate them. They have no options.

Russia and China did not even insist on voting on the matter. Everybody sheepishly moved on to the next item on the agenda.

The discussion at UN moved to the last ask from Russia – '*UN control over the project Mars. Russia should have a substantial say in the Mars Colonization Plan.*' As the Mars colonization plan was a topic of discussion, Mr. Leo Smith was invited by the UN and he was present in the room. Nobody expects him to agree to the demands of Russia. If Mickey's speech was any hint, Leo is about to actually rub it in. He has been invited by the UN to present his views and he is about to speak.

Leo Smith is 6 feet 2 inches tall. He is very good-looking, with sharp features. He is not only the richest person on earth he is also the most good-looking billionaire in the world. In his early days, when he was younger, he used to be lean, spectacled, and with a receding hairline. He used to look more like a geeky kid than a rock star he is today. He has changed everything about his looks. Got a hair transplant done, got rid of the spectacles, and has built a muscular body. It seems he spends a lot of time in the gym. He has consciously changed his image. He has bad memories of his school days when he was bullied a lot. He is not the same Leo anymore. Nobody can bully him now. Now he is the bully.

He can get into a fight with you on social media platforms without any instigation. He keeps instigating people to get into a fight on social media. He is infamous for challenging a fellow social media tycoon to get into a cage fight. And he was dead serious about it. He is not a person who will back out or get bullied. He is not a person to mess around with. Like Mickey, he also does not believe in holding back punches.

He is not only a bully but also a big bull of the stock market. He loves to talk up the prices of his company shares. If somebody tries to short his company stocks, he makes sure they pay for it. He single-handedly takes on a group of bears and defeats them in the duel. He will top it up by ridiculing them on social media. Leo launched limited-edition red satin short shorts in July 2020 to taunt short-sellers who were trying to take advantage of COVID-19 to short his stocks. Even Cyrus once tried to short Leo's company stock and Leo made sure to teach him a lesson.

Leo is also a piped piper, and his social media platform Y is his pipe. To get complete control of his pipe he bought it at twice the market price. Experts thought that Leo had overpaid and he would lose money on the deal. However, with the skills with which he played the pipe during the election campaign of Mickey, he managed manyfold returns. Leo bought the platform because Mickey was banned from the platform, he took it over, rebranded it, and then made it an in-house campaign tool for Mickey.

Leo has an unreal fan following. He has the capability of overnight doubling the price of any bullshit investment scheme, even random crypto named after a breed of dog, with a one-word social media post. His fans lap up whatever he says or even hints. For them, he is a God. His fans will follow him to the end of the world and beyond. With the Mars colonization project, he is about to prove that literally. Leo did not waste a minute and directly came to the point.

"My company SpaceY has been working on the Mars Colonization project for more than a decade. Like Noah's Ark, it is Leo's Ark and I only get to decide who gets to board

the Ark and who does not. You cannot bully your way to the Ark, you are free to build your own Ark."

Leo is living up to his reputation of being a bully. Leo is living up to his reputation of believing himself to be the incarnation of God.

"In my speech on 20th January 2025, during the coronation of Mickey, I promised two things.

1. *We will colonize Mars and provide each citizen of the Earth, who is more than 65 years of age and is scheduled to be switched off, the option to fly to Mars on my Ark.*
2. *We will provide the one-way ticket to Mars FREE of cost.*

I am not a person who will go back on my promise. But I am also not a person who will tolerate ungrateful people like Mr. Turin. It's in your interest and interest of your people that you accept our generous offer with gratitude and do not try to waste our time with your bullshit demands. You are in no position to make demands.

I have certain demands, that I expect all the members of the United Nations to agree to today. Time is of the essence and my offer of a free ride to Mars is not unconditional.

1. *Nobody, will interfere with Project Mars. My team will run the show and everybody else will follow instructions.*
2. *We will be required to set up base stations in all countries. All nations will give top priority to the project. Any delay at your end will be your*

responsibility. Any resources requested by my team should be given the highest priority.

3. *SpaceY team will have exclusive rights to fly and land on Mars. No other country or company would try to land on Mars as it can impact our project. Anyway, no other company is technologically anywhere close to land on Mars. If you want to land in space, please find a different planet.*
4. *SpaceY will have exclusive rights over the resources of Mars. SpaceY would be mining the Mars surface and during return flights bring back minerals from Mars. The cost of the free flights would be recovered from mining activity. Hence, it is paramount that SpaceY should have exclusive rights over the resources of Mars."*

Everybody in the room knows that Leo is bullying them and they have no other option but to agree to his demands. Right now, Leo's Ark is the only hope for people scheduled to be switched off in a matter of less than 5 years. A few days back Leo was a villain, now he is Noah. He is God, he is the only hope. And he loves to play God.

Earlier, the democratically elected leaders of various nations were under pressure to oppose the project Auto-off @ 65. They feared losing elections if they supported the project or if they were seen anywhere close to the conspirators. Now the situation is just the opposite. People across the world have digested the fact that the project cannot be switched off or reversed and 1st January 2030 is the doomsday. Now, everybody is concerned about their ticket to Mars and there is only one person who can save their life by giving them that ticket and that is Leo. Hence, now every world leader in the world wants to get close to Leo. Want to be seen close to Leo. Now, leaders around the world are under pressure from the citizens of their own country to support the project!

Chapter 4:
Preparing for Doomsday

20th January 2029 – The new president of the USA

4 years have passed since the big announcement of Project Auto-off @ 65. The second and final term of Mickey Cash, the 51st President of the United States of America has come to an end. The US Constitution allows only 2 terms for any president. Mickey wanted to do a Turin. He wanted to change the constitution to allow himself the third term. He argued that the world is staring at doomsday and that the USA and the world need the continuation of strong leadership. It tried to move the motion in both houses of Congress but a constitutional change required a two-thirds majority, and his proposal was defeated.

Leo was also very keen to become the president. Considering his fan-following, his God-like status, and the desperation among people to jump on his Ark, he could have easily won the election. However, there was a technical problem. Leo is a first-generation migrant from South Africa. He is a citizen of the USA but not a US-born citizen. As per the USA's constitution, only a US-born citizen can become the President of the USA, a first-generation migrant cannot become the President of the USA. It again would have required constitutional change and again would have required a two-thirds majority. Considering Leo's aura and charisma, he might have been able to gather the required two-thirds majority. However, first, the constitutional change and then the election process, together is a very long drawn process and requires a lot of personal time from the candidate. Considering the criticality of the Mars colonization project timelines, everybody including Mickey and Cyrus requested Leo to focus on the Mars project and not contest for the post of the President of USA.

At the same time, the three conspirators were not in the mood to give control of the most powerful chair in the world to somebody else at this critical juncture. Micky and the team have been able to bully the likes of Mr. Turin of Russia and Mr. Lian of China with the help of this chair. They do not want to let it go at this juncture. Hence, they decided to field the daughter of Mickey, Jane Cash.

Jane Cash was not a political heavyweight and could not have won the election on her own. She was very active along with Mickey during the first term of his Presidency but was somewhat withdrawn after he lost the 2020 elections. Jane was effectively a dummy candidate, the person who would really call the shots would be Mickey only, but only till 1st January 2030. Anybody would have expected that the co-conspirators or any of their protégée would never win the elections after the massacre they initiated with the project Auto-off @ 65. But that is history now. Public memory is very fickle. For a few months, these conspirators were villains, now they are the heroes. Earlier they were the ones who were killing people, now they are the only ones who can save these people by successfully executing the colonization of Mars. You can hate them, you can dislike them, but you can't ignore them. They are the only ones who are visible everywhere. Any other candidate had no chance. As Jane was not one of the conspirators, also helped her win the election. The conspirators openly supported Jane, but still, people were fine with voting for her. Another thing that worked in

Jane's favour was her age. She is only 47 years old. As human life is now restricted to 65 years, the Senate has passed a constitutional amendment that the incoming president should not be more than 60 years of age at the time of swearing-in. Most of the senior politicians in the USA were disqualified after this amendment.

Elsewhere in the world, the political environment is equally dynamic. Mr. Turin of Russia and Mr. Lian of China have ended up diminishing their aura of being strongmen and bullies of the world. They tried to bully the USA and the conspirators of the project Auto-off @ 65 but it backfired. The project was unstoppable and Leo's Mars option was the only option for survival. The citizens of Russia and China also understood the reality on the ground and started supporting Leo and hoped that Mars colonization would be a success. To save face and to appear not being bullied by the blackmailers, they announced their old mission to Venus and Jupiter. However, heart in heart they knew that any such project was not possible in such a short duration of 5 years. Still, the project made sense as even after the doomsday, every year a large number of people will get switched off at the age of 65 and would need an exit option from the earth. Hence China, Russia, and a few other countries joined hands to parallelly create their own space missions. At the same time, they did not opt out of the SpaceY offer of free transportation to Mars. They kept supporting the project, coordinated the bookings for their respective countries, and also provided all possible technical and other help to the SpaceY team. They also kept trying to enrol a few of their own people on the project so that over time they could learn from the SpaceY project and expedite their national project.

India also started investing in its national space mission project. They also managed to convince the USA and Leo for an exception to land ISRO rockets on Mars. The exception was agreed to by Leo on the grounds that India already had a very advanced project targeted at Mars and they had not really opposed Project Auto-off @ 65 since the beginning. However, they were allowed to land on Mars only after they agreed to two conditions. First, India cannot land their spacecraft to Mars before 2031 and secondly, mining rights at Mars will remain exclusively with SpaceY and India will never get involved in mining activities on Mars.

Politically, India is about to witness a regime change after a long period of 15 years. Mr. Mehta is completing his third term, his fifteen years in office in May 2029. He has decided to not contest for the post of Prime Minister again. Anyway, he is also scheduled to get switched off on 1st January 2030. It does not make sense for him to become PM for 7 months. After his speech in January 2025, post the project Auto-off @ 65 announcement, there was speculation that he would retire at the age of 75. He even offered to retire at 75, however, his supporters, party members, and allies were quick to convince him to complete the term and stay till May 2029. His friend Mickey also convinced Mr. Mehta to continue. Everybody was of the opinion that this project has already led to a lot of uncertainty and volatility in the world, at this stage India and the world cannot afford more uncertainty. Mr. Mehta has an image of being a proponent of peace and a mature leader. Many nations and their leaders look up to Mr. Mehta to guide them and uphold the interests of smaller nations. At this juncture world did not want to lose a mature leader who was not a bully and was strong enough to not get bullied.

Mr. Mehta did not agree to continue beyond May 2029 and suggested the name of Suryanath as his successor to lead the party and the country. Mr. Suryanath is a three-time Chief Minister of one of the biggest states in the country. With his vision, hard work, dedication, and strong-man tactics he has been able to develop the state at a very fast pace. Earlier, this state used to be one of the most backward states with major law and order problems. He is affectionately called Jogi by his followers as he lives the life of a monk. Now even the media and even his critics have started calling him Jogi. What also worked in favour of Jogi is that he will be only 57 years old in May 2029. After the announcement of the Project, similar to the USA, the Indian parliament also made a constitutional amendment to disallow politicians above the age of 60 to stand in elections. Many of Jogi's competitors in the party are already past 60 and will be forced to retire. Mr. Mehta also decided to retire in May 2029 and become a *marg darshak* or an advisor to Jogi and the party.

Mars colonization update

Leo had committed the year 2026 as the year for the unmanned mission to Mars and 2028 as the year for the first manned mission to Mars. Leo has a reputation for achieving the impossible and meeting all his promises, no matter how impossible the project looks at the beginning. However, he also has a track record of giving out aggressive timelines and during execution, he generally ends up slipping by a couple of years. With the Mars colonization project, he does not have that kind of luxury. If he slips by a couple of years, almost 700 million people will get switched off.

Thanks to US government support, availability of unlimited resources and cooperation of countries across the world Leo achieved the impossible this time a couple of years ahead of schedule. This was also possible because, for the first time, Leo 100% focused on a single project. Not 100%, but almost. Otherwise, he has a reputation for picking up more projects than it is humanly possible to handle, all at the same time.

Leo managed to successfully send the unmanned starships to Mars in January 2026, ahead of schedule. With the successful landing of the first starship, he started sending scores of unmanned starships to Mars every month. Almost every day one starship flies to Mars. The unmanned starships carried onboard Robots who were tasked to colonize Mars. They specifically had three tasks:

a. Mine the surface of Mars and bring back useful minerals including rare earth, Lithium, and other available minerals which were critical for batteries, computer chips, and many other aspects of the project.

b. Prepare the Mars surface and environment for agriculture through various artificial means.

c. Prepare the planet Mars for human habitation.

For two full years, not a single human being was sent to Mars due to two reasons – one, the travel to Mars was not humane, travel time itself was 9 months. Second, it was almost impossible to survive on Mars without key life requirements like water, air, food, etc.

Over 2 years the SpaceY team managed to not only reduce the travel time to Mars from 9 months to 4 months but also managed to start some basic agriculture on Mars. The first manned starship flew from Earth on 27th January 2028 and reached around 1st June 2028. On board, this aircraft were 24 dedicated employees of SpaceY who were well trained for this inhumane travel and inhumane Mars environment. These were young people in their twenties and thirties. They were healthy and trained. They have been training for this for more than 3 years.

The first manned mission to Mars was a 100% success. All 24 astronauts survived the trip and the Mars environment. They initiated the process of colonization of Mars. With the success of the manned mission, between June and September 30th, 2028, SpaceY sent another 20 starships to Mars with around 1000 human beings, all trained for the purpose. All these 20 starships were launched to 20 different locations on Mars and they were supposed to create 20 different colonies.

All these launches were successful and hence commenced the successful colonization of Mars.

Gung-ho with the successful landing of the first 1000 humans on Mars and their successful survival in Mars's harsh environment for more than 30 days, SpaceY activated the process of sending senior citizens to Mars. People who are about to be switched off.

SpaceY opened up the bookings for the starships for senior citizens on 1st July 2028, after the 30-day survival target was successfully met for the first set of space travellers. Initially, bookings were opened for 10 aircraft every day carrying 1000 people each. The bookings were opened for the period between 1st January 2029 and 31st December 2029. The booking was on a first come first serve basis and it got filling for the month of October 2029 to December 2029 in a jiffy. It took less than 15 minutes for the seats to get booked. However, for the period earlier than October there was stiff resistance. People wanted to fly off as close to the switch-off date as possible.

Leo was forced to do a public speech on television. He was very short and to the point.

"SpaceY has promised one-way transportation to Mars. The bookings are open and it's time you book the seats. It has been brought to my notice that people are insisting on seats for dates closer to the end of 2029. Please note, that it is not technically possible to fly everybody in one go and hence we have to spread it out. Book now otherwise you won't even get those seats. I never promised everybody a seat on the last flight.

If even one seat goes unutilized then I would believe that supply has met the demand and I have fulfilled my promise."

The message was clear. There was some resistance but people knew they did not have many options. Grudgingly, they booked whatever seats were available, and within 24 hours, all the seats for the planned flight in 2029 were booked.

There were some cases where younger people wanted to accompany their parents to Mars. Few youngsters were not comfortable leaving their parents alone. As it was not possible for the old parents to survive the switch-off, their children decided to accompany them to Mars. However, SpaceY denied any tickets to people younger than 64. SpaceY clarified, that a limited number of seats are available and they are running against time to provide a sufficient number of seats for senior citizens. SpaceY made a social media announcement specifying that travel for younger people will only begin in 2030 after the switch-off date.

When the booking started the tickets were free, as promised. Hence some people started booking multiple seats to build options. To restrict multiple bookings which would lead to a wastage of seats later, SpaceY first tried to restrict only one booking per ID card. ID card was also required to verify the age of the passenger. However, there was no single unique ID card for citizens across the world. SpaceY tried to allow different ID cards for different countries but it had its unique challenges. There were ID cards where online and automated verification of the genuineness of ID cards was not possible. There were cases of fudging of ID cards. India's Aadhaar card was one of the best ID proofs across the world which could be automatically verified online and had biometric data of passengers to be verified again at the time of boarding.

SpaceY was forced to transfer the responsibility of bookings to respective countries. They started allowing quotas for each country based on the number of 65 years or older population. Respective governments were responsible for taking the bookings. Respective countries were responsible for submitting ID proof and biometric data of each passenger in advance to be verified at the time of boarding. No last-minute replacement or switching of passengers was allowed. Respective countries were given strict instructions to maintain a transparent and fair booking system. They were warned against any kind of favouritism.

To restrict multiple bookings and last-minute withdrawals, which would lead to a wastage of seats later, SpaceY was forced to ask for a security deposit. The security deposit amount was fixed at $2000 per passenger, much lower than the cost of actual transportation. It was more of a deterrent to ensure that seats were not wasted. Currently, each seat is worth one life. Security deposit was supposed to be encashed only if the passenger failed to turn up on the designated date of the flight. It was not possible for individual passengers to furnish security deposits as many of these people were very poor. Hence the responsibility of furnishing the security deposit was also transferred to the respective national government. After some negotiations, SpaceY agreed to take bank guarantees equivalent to 25% of the quota for each country. The expected failure rate or percentage of people not turning up on the designated date was not expected to exceed 25%.

There were reported cases of unfair allocation of seats in many countries. In many countries government officials tasked with the fair and transparent booking of seats invented ways to make money by playing around with the seat booking process. SpaceY team was too busy to get involved in the seat allocation process and ensure fair and transparent allocation. They kept issuing warnings to erring countries and at times penalized a few countries by reducing their quota allotment. SpaceY had not anticipated that seat allocation would become such a challenging task.

SpaceY opened up seat bookings for the period of October to December 2028. This time again there was massive resistance to go full one year in advance. People were confident that SpaceY would be adding more starships for 2029 as currently there is a huge gap between demand and supply. However, a single tweet by Leo on social media did the trick. People got in the line to book the ticket. Still, few seats went empty on these flights and Leo was incensed.

Leo took this opportunity to claim that he has fulfilled his promise of providing seats to everybody and he cannot be held responsible for not fulfilling his promise if anybody does not get a seat in future. He went on live television to make this announcement. He tweeted this several times on his social media platform Y. People hated him for this but at the same time people agreed that Leo has a point.

Then came the bad news. Many passengers from the early flights did not survive. The survival rate was less than 70% for the flight and less than 50% for people after landing in the first few days. All these passengers were over 65 years old and had received zero training. Also, there was no meaningful medical facility yet on Mars. Leo and the SpaceY team were expecting some casualties and were prepared for it. However, they did not expect the causality to be as high as 50%. However, Leo is not a personality who will feel guilt or remorse. He is half robot himself. There are rumours that he is regularly high on drugs, which ensures human feelings like guilt and remorse do not even touch him. Rumours, that Leo has reinforced this by releasing a video of himself smoking marijuana during a podcast. He again went live on his social media platform, Y, and made a public announcement.

"We would like to make people aware that we have noticed casualties during flight and on landing on Mars amongst the senior citizens. The casualties are as high as 50%. This is a very sad moment for us. To date, we had a 100% success record. On our earlier flights when our crew flew to Mars, we had not encountered a single casualty. This shows the casualties are related to age and health.

We request people to take special care of their health before the flight. We also request people with medical issues and people who are very old to reconsider whether they actually want to go to Mars. We are now making it compulsory for passengers to disclose their medical conditions at the time of booking and at the time of boarding. We will reserve the right to deny any passenger boarding if our team believes that the chances of survival of that person are very low. The flight to Mars and stay on Mars is not very easy. We do not have any doctors, hospitals, or medical facilities on Mars. If you are ill and have low survival chances, you might like to live till the end with your family member here on Earth and experience a painless switch-off rather than dying a painful death on Mars."

The message was short but the impact was enormous. Until now, people were seeing Mars as a sure-shot option to escape death. Now even the Mars option has a survival probability of less than 50%. People were surprised by the emotionless, guilt-free, blunt message by Leo and SpaceY. But Leo has this image. He considers himself as God. He will try many things and will fail many times. But will never feel guilty or get bogged down by it. He will stand up and do it again. Till he succeeds. He is not a person who will feel or show remorse. For him, anyway, these people were about to die. For him, it was their choice that they boarded the starship. They were fools if they were not aware of the risks. He had already made everybody sign a no-liability consent and declaration form to protect SpaceY and himself from any legal or other liability.

After people became aware that the survival rate was less than 50%, many people decided to opt out. As there was no option to cancel, change, or transfer the bookings there was a lot of hue and cry. SpaceY was forced to provide an option to cancel the booking. Cancellation was allowed up to 60 days in advance of the flight date. The option to change or transfer was not provided. Even after all the cancellations and huge drop in interest, supply was much lower than demand. Still, many people were desperate to book a seat on the starship and were not able to.

Laxman from India is now much more confused and perplexed. His father who was 75 years old at the time of the announcement has already left for the heavenly adobe leaving behind his mother who is now 73. His mother is still quite healthy. She maintains herself well, with a controlled diet and regular exercise. She is mentally also very strong.

Laxman was lucky enough to book tickets for her on the June 2029 fight along with her brothers and sisters. Laxman's mother has 9 siblings and almost all of them are now 65 plus. They have booked tickets together in the hope that they will be able to stay together on Mars and will be able to live the last few years of their life in each other's company. With the news of a low survival rate, his mother and her siblings are having second thoughts. Many of them think that it is better to stay 6 months extra on Earth and die a painless death rather than take chances of travelling to Mars and die a painful death in the process.

Laxman has not cancelled the tickets yet. Rebooking is not allowed after cancellation. Cancellation would mean giving up the option of Mars. A couple of her siblings, who are much older and physically not doing very well, have already opted out. They have cancelled their tickets.

On the other side, Laxman's in-laws, and Urmila's parents are also having second thoughts. They also were booked on the same flight. Urmila's father is now 89 years old. He has diabetes and a few other medical complications due to which he does not like to travel much. Travelling for four months in a spaceship is something out of the question for him, especially after hearing about the survival rate. Urmila's mother is also quite old and has hearing and visibility problems. She is fine both ways but she is not ready to go alone. She is okay to go only if her husband is also going. She does not want to live all by herself on Mars. Urmila and Laxman tried a lot to convince Urmila's father to fly to Mars but he had made up his mind. Reluctantly, Urmila cancelled the tickets for both of them.

Laxman finally managed to convince his mother and a couple of her siblings to fly to Mars. Laxman is himself not sure either way but he has to take a chance. Both the options are bad, but the option of not doing anything, the option of not trying to save Mother by allowing her not to board the flight to Mars is an option filled with guilt and regret.

Mr. Lee in China has a similar dilemma. His parents are around 65 years old but his grandparents have already crossed 85. Grandparents are not doing very well medically and their chances of surviving the gruesome flight to Mars are almost negligible. The whole family after deliberation decided to send only their immediate parents who are still decently fit and might be able to survive the ordeal.

Many countries started their internal screening process based on health. Priority of booking was given to people within the age bracket of 65 to 70 years. In most of the cases, people above 70 years of age were being cajoled to not apply. Still demand exceeded supply. There were more than 700 million people in the eligible age bracket. It was highly unlikely that at this rate, SpaceY would be able to transport even 20 million of these people.

To the credit of SpaceY, they were doing their best. They were performing much better in terms of ramp-up compared to their plans before project Auto-off @ 65 was initiated. They have plans to send only a few thousand people to Mars in the next 10 years. Now they are sending a few thousand every day. SpaceY kept ramping up aggressively. They kept adding new flights and opening up new seats for booking. The seats kept getting filled up within minutes of opening up.

India managed the booking of tickets for spacecraft to Mars very well compared to other countries. From day one, the booking window was integrated with a national unique identity system called – Aadhaar. Everybody was required to authenticate their age and self via an Aadhaar-based online verification system before they could make the booking. The whole process of verification and booking was online, automated, and without any human intervention. The system worked very smoothly and was very fair and transparent. India is the most populated country in the world and hence it needs a large number of seats. Enough seats were not there. Businessmen, politicians, and bureaucrats were trying to jump the queue and were applying all possible pressure. It is almost impossible to manage seats for everybody, whoever did not manage to get the seat in time, will get switched off. Getting the booking is a matter of life and death.

Mr. Mehta has indicated that he does not want to travel to Mars and would prefer to get switched off on D-day. Mickey and Cyrus have also indicated that they would not be travelling to Mars and would prefer to get switched off. Their supporters, fans, and near and dear ones are convincing them to change their minds. Leo himself has made a public request to them to fly to Mars. However, they haven't budged yet. Despite their reluctance to travel to Mars, Leo has on his own booked tickets for all three of them on one of the late December flights.

On the other hand, Mr. Lian and Mr. Turin have publicly maintained a stand that they have not booked their tickets yet. They have not articulated their position publicly but have directly or indirectly indicated that they would be able to build their national space missions in time and use their national spacecraft to migrate from Earth before D-day. However, they have neither abdicated their throne nor have they announced their successor. There are rumours that they have already booked their tickets on SpaceY's spacecraft. SpaceY has neither confirmed nor denied the rumours. They are keeping all passenger identity information confidential.

Green Origin

One of the surprising success stories when it comes to space missions is Green Origin, founded by Ted. Ted is 65 years old business tycoon. He is ranked number 2 in the world's richest list only after Leo. Both Leo and Ted have substantially increased their lead over the other rivals on the list due to their respective space missions.

Ted is also flamboyant, but not as omnipresent as Leo. Ted used to be a successful investment banker. He left his job in the early nineties and started an online book retail company called Bookzone. Over the period Bookzone became the biggest online retailer in the world. Bookzone also ventures into the cloud services business, which initially started as an in-house project to support the technical requirements of its own business. However, later they opened up their cloud infrastructure to the world and ended up also becoming the largest cloud infrastructure provider in the world.

After Ted turned 60, he retired from the executive role at these two companies and started focusing on his space mission in competition with Leo. Green Origin also managed several successful space missions as early as 2015, but they were never very visible in the Media. Unlike Leo, Ted is a more private person. He loves to live his life and owns the biggest yacht and big mansions, but is not as eye-catching as Leo. He is bald and has a drooping eye. He is not as good-looking as Leo. Ted, like other billionaires we discussed, also had an affair outside his marriage, and somehow pictures of his affair got leaked to the media leading to a very embarrassing situation for Ted. The leak also led to his divorce. Ted ended up giving a substantial part of his wealth, in the form of shares of his company, to his wife as part of the divorce settlement. Still, despite the payout, he continues to hold the title of the second richest person in the world.

Green Origin rocketed into the race to space and in space too, Ted managed to conquer the number two position. The world was betting that either India, China, or Russia would be number two in space and would be able to colonize another planet after Mars. But Ted managed to beat all of them. Ted decided to colonize the Moon instead of the planets. It was counterintuitive to colonize a satellite when everybody else was looking to colonize a planet. But it was a very smart move by Ted. Mankind landed on the moon more than half a decade ago hence lot of data was already available about the moon, its surface, and its environment. Secondly, the moon being closest to Earth, also has one of the best climates suited for human survival. The moon is almost the same distance from the Sun as Earth. Hence Moon is neither too cold nor too hot. Thirdly, the distance from the Earth to the Moon is much shorter than any other planet in the solar system. This makes travel easier, cheaper, and safer.

Green Origin rocketed into space travel and managed to colonize the Moon in 2029, one year later than when SpaceY colonized Mars. But most importantly, one year ahead of the D-day. And unlike Leo, Green Origin never promised anybody a free ticket to space. Green Origin kept the project very private and exclusive. It offered very select people an alternative exit plan from the Earth but for a price, a very steep price. It is rumoured that Mr. Turin of Russia and Mr. Lian of China are a few of the first customers and one of the biggest backers of Green Origin. They are rumoured to have backed the project after they were blackmailed and ridiculed at UN headquarters by Leo in July 2025. Green Origin has vehemently denied all such rumours, but under the given circumstances it is very easy to believe these rumours.

Ted's Green Origin is charging $ 10 million for each seat onboard his spacecraft. Even then, it has enough takers. He also is not able to meet the demand with supply even at such a steep price. And this price is for transportation only. He is charging extra for stay and facilities on Moon. There are many things that he is offering which even SpaceY is not offering. For starters, he is allowing youngsters, people as young as 10-year-old the option to fly to the moon. Secondly, he is allowing people to travel back from the Moon at their will. Third, he is allowing them to choose the facilities they want on the moon, of course, there are limited options, and all come at a steep price. But money is not a limitation for many people. And he needs only a handful of them. He is not making it a mass project, anyway Moon's land area is very small compared to Mars or Earth.

Unconfirmed reports suggest that Ted has pitched his space odyssey to the heads of states of many nations, business tycoons, rock stars, sports stars, movie stars, billionaires, and many other people who have a net worth of more than 100 million dollars. Till D-day, he has only 1000 seats available with him and it is believed that all the seats are sold out. People have started bidding for seats on his spaceship.

Ted is marketing his colony on the moon, which he has named TedLand, as a super-premium retirement home exclusive for the Richie rich. TedLand is marketing the Odessey as invite only opportunity to retire in heaven without dying. Although he is specifically targeting rich people who are about to be switched off on D-day, he is allowing younger ones to also board the spacecraft because some old billionaires want to migrate along with their families. One billionaire insisted on taking his dog along with him, and his request was accepted. Of course, the price for the dog's ticket was double compared to any other human being.

Chapter 5:
2030: Doomsday

The Doomsday – 1st January 2030

The doomsday has arrived and as expected, it is nothing less than a massacre. It is the biggest massacre in the history of mankind. Around 700 million people were above the cut-off age. Out of that, only 60 million people have successfully reached Mars. Another 40 million people have left the Earth and are on the way to Mars, the expected survival rate is only 50%. Around 50 million people have died trying to escape to Mars, some on the way and some after reaching Mars. Around 500 million people opted out and decided to get switched off without any pain. Around 50 million people did not manage to get a ticket to board the spacecraft to Mars. These are just rough numbers. Real numbers will take time to compute. Real numbers might never come out.

Earth is lighter by around 700 million people, around 10% of the world's population. Out of 700 million people, around 80 million people reached Mars, assuming a 50% survival rate of those in transit. Rest 620 million either died trying to migrate to Mars or got switched off.

As expected, there is gloom everywhere. The conspirators of the project Auto-off @ 65 gave a full 5 years to people to mentally prepare for this eventuality, but today people are still gloomy and miserable. Nobody can ever prepare enough for losing their loved ones. Nobody can prepare enough for losing their parents. There is a massive anger among people. Anger and helplessness. Anger against the conspirators of the project. They have run the full circle. They used to admire and worship these billionaires before the project was announced. Then they were angry with the creators of the project. Then they felt helpless against the blackmail of the conspirators. Then they were hopeful that they would help them with the exit option to Mars. And now they are again angry with the evil trio for killing so many people at the switch of a button.

Relatives of those who managed to board the spacecraft to Mars also have mixed feelings. Many of them have lost their loved ones in transit to Mars. Few were euphoric that their parents managed to reach Mars against the odds. But since then, they do not have much news about their parents and how they are doing there on Mars. SpaceY has just provided data on who managed to reach Mars and who did not. If somebody did not survive after reaching Mars, that data also was shared with their relatives. But beyond that, there is no further information. There is no information on what kind of facilities are there on Mars. There is no way to communicate with the relatives. The only hope for people is to pray to God or pray to the person who thinks himself to be God. The self-designated God announced that communication facilities would soon be available between Earth and Mars, but he did not provide any definite timelines.

Laxman spent four months in a lot of anxiety when his mother was in transit. He was elated when he got the confirmation that his mother had landed on Mars and was doing well. However, he was immediately very sad when he got the news that one of his mother's older siblings and his wife did not survive the ordeal. Laxman knew that his mother was very close to her sibling. He is well aware that she would be completely devastated. Laxman would have preferred to be next to her in such a moment of sorrow, but he is helpless.

Meanwhile, SpaceY also confirmed that Mickey Cash has successfully landed on Mars and he is doing well. However, SpaceY also shared the sad news that Cyrus Clark was not able to survive the tough flight to Mars and passed away on the way. Leo was able to convince the two conspirators at the last moment to take the flight. He convinced them that the world still needs their leadership. Mickey Cash, himself was also very keen to continue to lead the world. He believes that nobody can lead the world like he can, not even his daughter. He treats his daughter like a mask. He had been calling all the shots from behind even after the formal transfer of power to Jane Cash had happened. Mickey Cash wants to continue to guide and direct his daughter from Mars.

Leo was able to convince his co-conspirators but he was not able to convince Mr. Mehta, ex-Prime Minister of India. Leo tried very hard, he travelled to India twice to convince him, but Mr. Mehta did the budge. He had two arguments, one, he had promised his people that he would be the first one to die, and he would not go back on his promise. He quoted some ancient Hindu text *'Raghukul reet saada chaali aaye, praan jahe par vaachan na jahe.'* Leo did not understand what it meant but he was made to understand that loosely it means that once you have made a promise, you should stand by your promise even if that costs your life. Leo argued against it. He gave various logic including that since the citizens have been given the option to travel to Mars, technically Mr. Mehta is not going back on his promise, but Mr. Mehta was adamant.

The second argument that Mr. Mehta gave was even more difficult to argue against. Mr. Mehta argued that the whole premise of the creation of Project Auto-off @ 65 was that old people should get out of the way and relinquish control to younger people. If Mr. Mehta continues to call the shots from behind the curtains, like Mickey Cash is doing, then the creators of the project themselves are not abiding by the core principle on which the project was conceived. However, Leo differed with Mr. Mehta also on this point. He argued that the general laws apply to lesser mortals, rules do not apply to them. They are exceptions, even science, and God's created world always have exceptions for each rule. Leo's arguments worked with Mickey and Cyrus, but Mr. Mehta was an exception. Exception to the general human nature of twisting the logic to fit your needs. He stuck to his words and decided to get switched off rather than board the flight to Mars.

Even after ceding the post of Prime Minister to Jogi in May 2029, Mr. Mehta has ensured that he gave advice and direction to the new Prime Minister only when it was asked for. On one of two occasions, he gave his advice on his own but only as advice and not as a direction. It was very different from the way Mickey Cash had been operating after ceding power to Jane Cash. He has been running the show from behind the screens, in fact not even behind the screen. He was brash about it. In many national and international meetings, he would attend himself and do most of the talking and Jane would remain a mere spectator. Jane was the pretty face mask; Mickey was the one in control. Mickey was one of the creators of the project which aimed to transfer the power from the hands of senior citizens to the younger generation, but Mickey himself was not practising what they had been preaching.

Giving up power does not come easy to successful men. World leaders across the world, with very few exceptions, tried to control the government from behind the screen even after they had to cede power due to age. Many of them continued to sit on the throne till the last day before they boarded the spacecraft. Mr. Turin and Mr. Lian of China were no exception. They continued to hold power till they boarded the spacecraft. Till the last moment, there was no clarity on whether they will get switched off or they will board the spacecraft to Mars. As expected, they did not choose to get switched off. However, they did not board the spacecraft to Mars. They boarded the spacecraft to TedLand.

When SpaceY confirmed that Mr. Turin and Mr. Lian had not boarded one of their spacecrafts, it came to light that they had decided to board the spacecraft to the Moon, to TedLand. People are speculating whether Mr. Turin and Mr. Lian paid $10 Million each for their seats or whether Ted offered them free seats considering their position. The secret would never get out but people are more or less convinced that they paid for it. Ted is offering a lot more facilities on TedLand including the facility to communicate with Earth, obviously, for a price. On the other hand, SpaceY is providing very basic facilities on Mars in comparison and communication with Earth is not one of them. It is feared that Mr. Turin and Mr. Lian would continue to wield power and control the politics of their respective countries, not from behind the screens but from the moon.

Doomsday has just arrived, how things would pan out further is everybody's guess and would be clear only over some time. Till now everybody, including Leo and SpaceY was only focused on getting more and more people to Mars. To fulfil his promise of giving an option to every senior citizen scheduled to be switched off option to exit Earth. He did not fulfil his promise completely, but he did not do badly either.

He, in any case, does not feel that he did badly on his promise and does not feel guilty about it. He feels proud that he managed to transport more than 60 million people to the moon in such a short time. And we have to give it to him, this is not a small achievement. No other mortal has been able to achieve even one percent of that, including Ted.

Now that D-Day has arrived and most of the senior citizens have been switched off, the pressure for seats has eased. The backlog has been cleared and now demand would be only from people who are turning 65 over time. SpaceY has ramped up so much in the last year that managing future demand should not be a problem anymore. In a way, D-day has arrived and it is also gone. It is now a matter of history. The world is still there, we may feel gloom and doom but the world is still there. Mankind still exists and now exists at 3 places in the solar system against only one place a couple of years ago.

Still, people are feeling gloomy, people are angry about having lost their loved one. People want to crucify the creators of the Project Auto-off @ 65. However, they don't know how to do that. Cyrus Clark has already left for his heavenly abode, and Mickey Cash has left the Earth and migrated to Mars. The only person they can target their anger is Leo, who is still there on Earth. Hence, everybody including the public, media, and political parties started targeting Leo. There was a demand for the UN and International Criminal Court (ICC) to declare him as a mass murderer, and for his 'crime against humanity' he should be arrested and hanged. The chatter and demand for crucifixions was everywhere - social media, print and television media, parliaments, everywhere. Even on Leo's social media platform Y. Leo manipulated the algorithm of Y to ensure only positive comments about him are visible and negative comments are buried, but he was not successful. The demand was everywhere and too intense to be contained by algorithm manipulation. Hence, Leo decided to counter it by delivering a live speech. He believed he had done more good than harm and he needed to give the right perspective to the people.

Leo, now 58, still has 7 years before he gets switched off himself or is forced to migrate to Mars. He still looks like a rock star. He is still very fit, good-looking, and can take on anybody in a cage fight. The speech he is about to deliver is no less than a cage fight. The only difference is, he is all alone against the whole world. In the past he had his co-conspirators standing next to him. However, Leo is not a person who needs anybody to stand next to him. He has the confidence to take on the whole world, all on his own. He is the big bull who can eat a pack of bears for breakfast. He is the big bully who can bully powerful dictators like Turin and Lian. Now, even those powerful opponents are no longer there to oppose him. He believes he can easily stand up to the challenge and change the public's perspective with one speech of his. After all, he is the God, at least he believes he is no less than a God, and with Mars mission, he has proved it.

"Dear fellow citizens of the Earth and the citizens of Mars."

People are wondering, why he is addressing to citizens of Mars, is the speech being telecasted on Mars as well? Leo was quick to provide clarity.

"This speech is being telecasted not only on Earth but on Mars also. First of all, let's congratulate ourselves for being

able to successfully colonize Mars. Together we have achieved the impossible, we have not only saved mankind but also taken a big step toward making it prosper beyond Earth, across the solar system.

Mars Mission, the mission to colonize Mars, was conceptualized by me around 20 years ago, to provide mankind an alternative. The mankind had all its eggs in one basket – the Earth. Any risk to Earth was a risk to mankind. Mankind has survived several such extinction possible events including asteroid collision. We were lucky to survive. But dinosaurs did not. It would have been foolish to trust our luck forever. There are many extinction-threatening events on the horizon including global warming, pollution, and the threat of extraterrestrial attacks. The basic law of risk management stipulates to always have a plan B. We mankind, did not have any plan B before SpaceY stepped in. Now we have a real plan B. I congratulate my team, the SpaceY team for making this possible."

People listening to him were not able to disagree with him on the point that it is a big achievement that mankind now has a plan B. If something happened to Earth, at least we have an alternative option. If the Earth is suddenly destroyed, mankind will continue to survive and grow due to the existence of people on Mars.

However, people are not truly concerned about the plan B right now. Risk to Earth is something for intellectuals, common people never even consider those things. What they care about is themselves and their loved ones. They do not care about mankind or Earth which people like Leo talk about. They are infuriated because they have lost their loved ones. To mitigate a risk, that is not even on their horizon, they were not willing to make sacrifices today. Leo might feel that it was a small price to pay, but people do not look at it as a small price to pay. For them, it was a genocide.

"I know what you, many of you are thinking. You people do not even have a vision to understand that mankind is at risk, most people are like ostriches, they bury their heads in the sand when they see risk and think the risk will go away. That's why you people need a leader like me to protect you. That's why you need to trust your leaders. If you are not intelligent enough to protect yourselves, as a minimum, trust the people who are intelligent, visionary, and committed to protecting you. Do not bite the hand that protects you."

People are infuriated, they are thinking - first, he killed our parents and now he is insulting us, calling us names, calling us fools. They are not able to believe his audacity, if he was in front of them, they would have killed him. But he is there on screen and not there in person. Hence there is very little they can do right now but to listen to him.

"I know you people are not able to appreciate what I have done for you people by providing option B, by colonizing Mars. You will appreciate that when some extinction-threatening event happens. An alien attack, asteroid collision, you never know what that event would be and that's why you cannot plan just to take the problem head-on. You also need a defensive plan; you need a plan B. Till that time you can continue to bury your heads in sand and keep cursing me for the cost paid for plan B.

I can understand, not appreciate, but understand, you people do not have the vision to anticipate the possibility of such an extinction event. What I am not able to understand is you people are not even being able to see what you people have done to Earth in the last 100 years. You people with your selfish act have endangered Earth. Pollution, global warming, and worsening climate are all man-made problems. You blame me for the switch-off, I blame you for all the deaths due to pollution – air pollution, water pollution, noise pollution, and pollution of the Earth's surface. You people are killing millions every day with your pollution and are not willing to accept the blame. You want to blame others; you will shift blame on each other but would not accept you are also one of the culprits. You want to blame me for switching off the senior citizens. I am blaming you for forcing us to switch them off or send them to Mars."

The listeners are now bewildered. They are not able to understand how Leo can have the audacity to shift the blame.

"The Earth is dying and you people are killing it. With eight billion people together conspiring to kill Earth, the Earth is alone and not able to defend itself. The stakes are stacked against Earth in the fight. I have reduced 10% of the people from the Earth to give Earth a fair chance. Still, the stakes are stacked against the Earth. You people need to give Earth a fair chance. You people need to stop killing Earth. You people have a habit of killing the hand that feeds you."

Now some of the people are able to understand what Leo is saying. Some percentage of the population, a small percentage, do appreciate the problem of pollution and global warming.

"You fools, who among you are foolish enough to believe that rising pollution is not a threat to Earth. Who among you is foolish enough to believe we can continue like this forever

without any implication? Global warming is real! You people are like the frog who cannot feel the warming water. You people are like the frogs who can feel the warming water but are too lazy to act. You people are like that frog who starts to enjoy the warming water and goes off to sleep. By the time you realize, you will be too weak to react. By the time you realize, it would be too late. Earth is the bowl and the atmosphere is the warming water. You were not jumping out yourself and hence, I was forced to throw 10% of you out."

Few people have now started appreciating what Leo is saying. They have also started feeling guilty too. But most of the people still want to crucify him. They do not like the way he is brazenly insulting them. They are not sure why they are infuriated with him now, for the reason he switched off their parents or because he is shamelessly insulting them.

"Tell me which fool here is not able to appreciate that with passing generations our genes are getting weaker. Humans are being born with diseases that never used to exist. There are so many kids who are born with eyesight issues. There are so many people who are born with cancer. We are drinking polluted water, we are breathing poisonous gases, and we are eating plastic. We are feeding plastic to our seas, fishes, birds, and animals. We are not destroying the Earth for ourselves but also for other living beings living on the Earth. How many species are going extinct every year? Who is to blame for it? You are to be blamed for it, and if left to me, I would have crucified you for that. You want to crucify me; I want to crucify you. I am not alone, many species of plants, birds, and animals want to crucify you for what you are doing to Mother Earth. You do not deserve Mother Earth. You people deserve to be kicked out of the Earth."

People have started to understand what Leo is saying, percentage of people feeling guilt is increasing in the crowd.

"I could have randomly kicked out 10% of the population, but I strategically and surgically pushed out the oldest people. You should be grateful to me for being so surgical. I could have kicked out any of you. That would have been the easiest thing to do. But we choose the hard path. We, the so-called conspirators of this project Auto-off @ 65, worked very hard for the last 10 years to specifically weed out the older people, people where gene mutation is higher. You people should be thankful to us. By weeding out the old people we have given you people a chance to survive and prosper. We have tried to protect the human genes from degenerating at a fast pace. We have tried to give fair share of Earth resources to younger people, to you people. We were forced to snatch the resources of the Earth from the older people in your interest. They were monopolizing the resources. Now you have a fair chance. Look around you, till 5 years back, most heads of the nations across the world were these old people. All big businessmen and all major politicians were these old people. The world was in control of old people. If the control is in the hands of old people, they will think only about old people. They will think about how they can live longer and hold on to power. They will not think about the next generation. I have kicked out dictators like Turin of Russia and Lian of China from the face of the earth. Shouldn't you be thankful to me for that? Shouldn't you be grateful to the creators of the project? And what are you people doing? You people are rallying against us, the people who worked in your interest, people who sacrificed their lives also for you people, you are blaming them? You should

forget the Gods you have been worshipping to date and worship the creators of this project."

People can appreciate that this project has led to the kicking out of many dictators who were sitting on their chairs for years and were impossible to get rid of. Turin and Lian and just two big examples, there are several of them across the world. Not only country heads but heads of companies, heads of households, heads of associations, heads of communities, and heads of religions. This project has helped in the change of guard and transfer of power in many organizations. For good or bad, only time will tell but with 65 as the cutoff date, it would be impossible for any dictator to last long.

"I am no God, but I am not like you people, you filthy human beings who are killing the Earth. I know you people do not understand your self-interest. You people do not understand the language of love and care. We have been giving you love and you have been giving us hate. We have been working for you and you have been looking to crucify us. You people only understand the language of threat. You people can be brought in line and made to act only by threat. When we beg you, request you, educate you, you do not understand. We tried to beg, request, and educate you about global warming, but you did not care a damn. I was forced to kick out 10% of the population from the face of the earth. I warn you now, I threaten you now if you people do not mend your ways and start taking care of Earth if you people do not stop polluting Earth, if you people do not stop killing Earth, I will kill you. I will keep kicking out more and more people from the face of the Earth if the situation does not improve. The earth is not your birthright. The earth does not belong only to you. You are sharing it with other living beings. BEHAVE. STOP

THE POLLUTION. USE THE EARTH RESOURCES JUDICIOUSLY."

Now the final part of the message was received and understood by everybody. They might have not understood the global warming part, but they understood the threat given by Leo that he could kick out more people from the face of the earth. There is no reason to take his threat lightly, he has just kicked out 700 million people from the face of the earth and he is very capable of doing it again. Leo is playing God; he loves playing God. Leo is right in saying we humans do not respond to love and care. But we humans respond to threats and fear. Leo has instilled fear in people. Not fear of God, but fear of Leo.

Doomsday – Part 2

The world has been preparing for this doomsday. If only you can prepare enough for such a day. Most people have taken a holiday and were spending time with their parents who were about to be switched off. The government and people have made arrangements for mass funerals. It was not possible to prepare for individual funerals for everybody. Families wanted to give individual funerals to their parents, as a mark of respect, but it was impossible to do so. With more than 600 million people dying on the same day, there was no way to have infrastructure for so many individual funerals.

Organizations have also been preparing for this D-day. Most organization changed their constitution to allow only people less than 60 years old to take up management posts in their organizations. This triggered a major change of power in the world. Corporations got rid of their board members and transformed into younger boards. Companies replaced many of their top management and key personnel and replaced them with younger people. Most Fortune 500 companies had many senior citizens on their boards and also a few in top management. They were forced to opt for younger people. Even in those organizations where key people were just below 60 years old or were still some time away from the cut-off age, as part of the transition planning, they started appointing younger leaders so that they could perform in the role for a longer time.

The biggest change was in political parties. Politicians across the world never had a retirement date. Most other professions had a retirement date, not politicians and to some extent businessmen. And these are the two most powerful sections of society and have the biggest influence on decision-making. Powerful people make rules for others but create exceptions for themselves and hence politicians have done the same. Project Auto-off @ 65 completely dismantled the hold of senior citizens over political parties and also the world's politics.

These old politicians were forced to cede control to the younger generation. This helped comparatively better-educated politicians make their mark in politics. It is still early days, but this is expected to have major ramifications. The ramifications are already very high in countries like China and Russia where the autocratic dictators have been holding the top post for decades. This project has liberated these countries from the clutches of these leaders. Whether these countries can get better leaders and more democratic government is a question on everybody's mind and would be answered only over time.

Change is visible everywhere, even in the smallest of organizations. Even a housing society which generally used to have retired people in their management committee now has younger people in the committee. Sports associations, clubs, associations of doctors, trade, and business associations, everywhere the hegemony of senior citizens has been broken and transition of power is happening to younger people.

TedLand

Ted of Green Origin also turned 65 in 2029. Hence, he was also supposed to get switched off on D-day. Ted did not board the spacecraft to Mars and build his private paradise on the moon, named after himself – TedLand. Ted is not as flamboyant as Leo but he is also full of himself. Leo and Ted were rivals not only for the world's richest person position but also in the print media business and space program. Although Leo and Ted were not rivals in their core businesses, they had different political preferences. While Leo was openly supporting Mickey for the post of President, Ted was supporting the opposition candidate. But he was much more secretive about it. Unlike Leo, he did not want to take an all-in bet and regret it later if the other candidate won. With a business like Bookzone, which is already unpopular for destroying jobs in mom-and-pop stores across the world, he did not want to be in the bad books of any politician. He maintained good relations with all the politicians and powerful people across the world.

These politicians and powerful people also needed an exit plan when the project Auto-off @ 65 was announced. They did not want to rely completely on Leo and SpaceY for their survival. They wanted to hedge their bets. Hence many of them put their bets on Ted and Green Origin. They invested equity in Ted's project and ensured that Ted did not waste any time due to the limitation of resources. After the project Auto-off @ 65 was announced, there was a deluge of funding for Green Origin. Green Origin was earlier progressing at a healthy slow and steady pace. But the deluge helped Green

Origin to rocket ahead in the race to space. The smart choice of Moon also helped.

The philosophy behind TedLand was the opposite of the SpaceY project. SpaceY was building a mass project where they wanted to fit in as many people as possible, 700 million of them if possible. On the other hand, Ted was building an exclusive project for his rich and powerful friends.

On the D-day, everybody was talking about SpaceY's successful Mars colonization. But what went below the radar was the success of Ted's Moon colonization project.

Ted built a very premium home away from Earth for his friends and himself. He created amazing recreational facilities for the exclusive group on the Moon. He also provided many facilities that were not there on Mars. One of the most important one of them was communication with Earth. From Day one, residents of TedLand had the option to communicate with Earth via voice and video calls, of course, for a price. For the rich inhabitants of TedLand, money was not an issue. This communication channel empowered them to continue their influence over the Earth and its politics.

People like Mr. Turin of Russia and Mr. Lian of China tried to maintain their autocratic control through their nominees on Earth. For some time, the nominees played along, but they soon started working independently disregarding the instructions of their mentor. They were aware that these people had been kicked out of Earth and had no chance of returning. Whatever limited power they have now is through their concealed black money and their erstwhile followers. They were aware that their erstwhile followers would have their individual ambitions and have no reason to fear them now. Hence, the influence of these autocratic leaders vanished very soon. Even their concealed black money on Earth started to vanish as the trustees of that money started to break the trust.

Ted was smart enough to structure his control over Green Origin and his wealth in such a manner that his trustees do not backstab him. He distributed his control over Green Origin to several people including his family members via digital password-based controls. No person or group of persons had the complete picture or control over the project. He also had control over digital passwords which only he had access to and through which he could control his empire on Earth.

Leo's ego was very hurt when he realized that he was not the only person who could give people an exit option from Earth. He expected people like Turin and Lian to beg him for seats on a starship but Ted's Green Origin denied him the pleasure.

He never missed an opportunity to discredit the Green Origin project. He dismissed the project as an exclusive Oasis of politicians with black money. He laughed at the project for selecting the moon as the destination. Leo elucidated that he always had the option of selecting the moon. Moon was a low-hanging fruit but he did not select the moon because the moon was not a good plan B. The moon is attached to Earth as it rotates around the Earth. Any risk to Earth has a high chance of being a risk to the Moon too. But secretly he despised the project as it was much less complicated especially because of its proximity to Earth. The location had the advantage of both weather and distance.

Sannyasa in LeoWorld

Leo, soon after the D-day rechristened Mars at LeoWorld. Ted has renamed Moon as TedLand, he was not one to be left behind.

Vanity runs in the veins of these billionaires. However, LeoWorld was not at all majestic in its operations. LeoWorld was so humble and basic that it could be compared to a jail. Everything was white, grey, or black, black being Leo's favourite colour. Residents were provided with small white pods where they could sleep. They had small black screens to watch all kinds of OTT channels. On Mars, earth's intellectual property rights laws did not apply and hence Leo has provided all the content free to residents of Mars. Anyway, they do not have much else to do there. There was no other source of entertainment. For poor people who were struggling to get two square meals on Earth, LeoWorld is like a five-star retirement home. For the upper class, the facilities are too basic. SpaceY has focused all its resources on transporting people from Earth to Mars and did not focus much on the quality of life on Mars.

Two major jobs on Mars were mining and agriculture. Both these jobs were 100% managed by robots of SpaceY. Humans had no roles to play there. The third major job was the hospitality of millions of people who were arriving from Earth every day. Providing them with a place to stay, something to wear, food, and other basics of life. This also was managed almost 100% by robots. There were dispensers from where food was delivered by machines. It was a self-service environment. Humans were supposed to approach these dispensers and take whatever they needed. The options were very limited and very basic. For the first year, everything was free, Leo is still to decide how he wants to structure it in the future. Currently, everything is funded through mining activities on Mars.

Humans who have arrived on Mars in groups were better off as they have somebody to talk to, somebody to take care of. Those who did not have anybody made friends very soon as there were not many options to spend time with.

On LeoWorld, the colonies were spread all across the planet. There was an internal communication system on the planet that could be used to talk to each other but there was no transport infrastructure for humans to travel from one colony to another. Those are part of the future plans.

LeoWorld did not have any medical facility. No doctors, no medicines, no hospitals. It was the responsibility of people to stay healthy, and if they didn't, hard luck! As most of the residents were old, there were regular deaths. In such cases, the robots would come and dispose of the body with little fuss. As food and other facilities were very basic, it was difficult to live an unhealthy life there. Rules of LeoWorld enforced mandatory two hours of exercise. Residents were supposed to use one of the many channels telecasting programs on exercise. Their television screens doubled up as coaches to help them with equipment-free exercises like Yoga and others. During these two hours, other OTT channels were not available, leaving little choice to residents but to exercise. However, the rules were not aggressively enforced as the number of Leo's robots was limited and they had other more important jobs like mining, agriculture, and hospitality to look after. Hence law and order task was left to the people themselves. People were expected to maintain self-discipline.

Enterprising residents started forming friend circles to keep themselves occupied. People invented or created their own entertainment systems in the company of each other. OTT and the company of each other were the only two means to spend time. Life was very basic and simple. It was more communist, unlike the capitalist nature of Leo. TedLand was the opposite because they had the luxury of a limited number of residents and clientele who were loaded with money and paying well for the services. On the other hand, LeoWorld has the limitation of a large number of residents and none of them contributing a single dollar. LeoWorld was a great leveller. Everybody was equal here, no rich, no poor. The only distinction between people was who was rich in health and who was poor in health.

Laxman's mom is also there in one of the colonies here. She has for her company one of her siblings. They have also made a few friends here. They watch TV together, they eat together, they exercise together. They like the life here. It is simple but not bad. There is no other responsibility other than maintaining their individual health. They do not have to do cooking for anybody or take care of anybody. They do not have to worry about money or materialistic goods. The life is free of any stress. Life is too simple to create any kind of stress. Even the lust of life is not there now, after being separated from their loved ones. The mindset is very stoic. They have seen death very closely while travelling to Mars. They do not fear death anymore. If they die tomorrow, they won't regret it.

In a way, people are living the life of a hermit. In a way, they are living in *Sannyasa Ashram*. In a way, they have attained *Moksha* or Nirvana.

Chapter 6:
Few Months After Doomsday

Leo in LeoWorld – 1st July 2030

Six months have passed since the D-day. The world has started adjusting to the new reality. When we say world, means two worlds now. The original world on Earth and the world beyond Earth – LeoWorld and TedLand. Residents of Mars have started adjusting to Mars and its simple way of living but they are missing Earth and communication with Earth. Relatives on Earth are also desperate to communicate with Mars at least once and get a confirmation that their parents are fine. They want to ensure that their parents are actually on Mars and no trick has been played on them. There are rumours on Earth that Leo has switched off everybody and Mars was just a story he was selling to distract people from switching off. He created the distraction from the whole of mankind so that they do not focus on the upcoming death, the upcoming massacre.

Leo is under pressure to prove that genuinely Mars has been colonized and people who were transported to Mars are safe and still living. To take care of the concerns Leo decided to do it in his unique style. He decided to address people from Mars.

On the morning of 1st July 2030, suddenly an announcement was made that Leo would be addressing both worlds from LeoWorld. There was a lot of excitement on Earth with this announcement as they would be seeing some live feed of Mars for the first time.

"Dear people of the old world and new, this is Leo Smith speaking from LeoWorld! This is my first trip to Mars and I am so proud. When I conceptualized the colonization of Mars, I wanted to be the first one to step on Mars, but I was beaten to Mars by millions of others. Due to strict timelines of project Auto-off @ 65, I was forced to stay back on Earth to coordinate the scale of operations we saw over the last two years. The operations of transporting millions of people before the switch-off date were not an easy task, but we did it. Now we move to the next phase of colonization.

Before I do that, let me share some good news with residents of both worlds. From today we are opening up communication channels between Earth and LeoWorld. Now people in both worlds will be able to talk to each other. People of both worlds will be able to see each other. We would allow video calls. For each resident of LeoWorld 30 minutes of video call to Earth every week would be free of cost. Beyond that, it would be chargeable. The call charge rate will be one dollar per minute. The cost of the chargeable call will have to be paid by the caller on the Earth side as in LeoWorld we do not have the concept of money yet.

Now let me tell you my vision for the colonization of Mars. The colonization of Mars is not complete yet as we have only old people here. Currently, we do not have anybody of reproductive age here, this way mankind would not be able to sustain itself on Mars. Secondly, we only have humans here. We need to get the other species on Mars too to ensure it is actually colonized. Other species also need plan B.

We are now allowing people from the age of 15 to travel to LeoWorld. We are also allowing two-way travel to and from LeoWorld. The travel to LeoWorld and back would not be *free of cost. We would charge $100K for a return ticket. Current residents of LeoWorld who are already 65 plus would not be able to travel back to Earth as they would die on entering the Earth's environment, but people from Earth can come and visit them. Stay up to one week would be free for people coming to LeoWorld and the cost would be included in the ticket. Beyond that, it would be chargeable. We would be promoting a lot of tourism. For people who are coming to LeoWorld on their own for the long term would be charged $ 1 Million for a lifetime stay here.*

To safeguard the Earth from humans with deteriorating genes, we kicked out people above 65 from Earth. We brought them to Mars. But now Mars would be at risk of having the majority of the population with deteriorating genes. That is not good for any new planet. We need to correct that. Hence, we are now also announcing the closure of our free Sannyasa plan. Beyond 2035, SpaceY will not be providing free travel and stay at LeoWorld to people who are kicked out of Earth. People would have two choices – either get switched off or pay $1 million for a lifetime stay at LeoWorld. People will also have the option to relocate to other colonies in space. One has already come up on Moon and others are also expected in the next 5 years. We will also endeavour to make LeoWorld a base and colonize other planets. But that might take time. People who have already arrived in LeoWorld and who arrive under the Sannyasa plan up to 2035 will continue to enjoy a free stay for a lifetime."

People are zapped. Very few can afford $1 million. For most people, Leo was not leaving any option but to switch off at

65. In a way, they are not very surprised. They have stopped expecting any remorse, sympathy, or compassion from Leo.

"We are also announcing a special one-time offer for people in the age group 15 to 25 to migrate to LeoWorld free of cost. This offer would be open to only the first 1000 people. 500 males and 500 females. The people under this plan would be selected after extensive medical tests and tests of the quality of genes. These people would be LeoWorld's Adam and Eve. These people will take Mankind forward on LeoWorld. These people will give birth to babies on LeoWorld, the original LeoWorld-born inhabitants of LeoWorld.

Initially, these young people coming to LeoWorld would be kept in separate colonies from the existing colonies where old people lived. We do not want the good genes and bad genes to mix initially. We also want to let Adam and Eve evolve on Mars as Adam and Eve might have evolved on Earth. We will divide them into 10 different colonies and leave them to evolve on their own. We will study their evolution from a distance. We intend to let nature take its own course. Over period of time, the inhabitants of these colonies might get connected to other colonies and mix, but for the first few years, we want them to evolve naturally. That will also help us understand how mankind might have evolved on Earth, something which is still a mystery and something we are still very curious about.

We will also start the process of bringing other species – animals, birds, and plants to LeoWorld. Leo's Ark will do the job. We will ensure that we have enough samples from each species here on LeoWorld so that in the future all species have plan B. Many species are becoming extinct on Earth every day, we will try to get these species to Mars and save them from extinction. We will ensure they grow and multiply here.

I intend to move permanently to LeoWorld before 2035. I look forward to staying in this brave new world."

The God Industry

Back on earth, people are now adjusting to the new world. The new world without senior citizens. The new world without old people.

Over the last five years, people have been guessing, forecasting, and extrapolating what would happen after D-day. Many of those forecasts have come true but there are many surprises also.

One of the major surprises is the God Industry. God industry is one of the biggest industries but somehow it remains under the radar. People see, feel, and interact with this industry every day but somehow do not see it as an industry. Most of the people on earth are customers or followers of one of the major religions. This industry has three to four major religions which rule the world. There are other smaller ones with their niche clientele. Few startups also keep trying to enter the market, but very few survive. The big religions are very powerful and do not let new players enter the market. These religions have their branches across the world, most of these branches working in coordination with the headquarters. However, like other industries, some of these major religions have loose structures and have evolved differently in different parts of the world. They have leaders and these Godmen are called by various names including Priests and Pandits. These Godmen command different degrees of power based on the power of their corporation, their position in the hierarchy, their region of operation, and their own client base.

The God industry was majorly impacted by Project Auto-off @ 65. In 2025 itself, just by the announcement, before anything was triggered. The power of the big corporations was questioned. The big corporations controlling this industry were asked how few humans, three conspirators, were able to change some fundamental things like 'death' which falls completely under the domain of these major corporations. These corporations rule on the basis of the belief and trust of their clientele. Products around 'Fear of death' are one of the biggest revenue generators for this industry. These major corporates tried to handle this sudden change in market dynamics in different ways. Most tried to capitalize on the sudden heightened demand due to anxiety around the impending switch-off date. They minted money like anything in the 5 years leading up to D-day. Most players in this industry ramped up their production capacity at a very fast rate to meet the sudden spurt in demand. They opened new branches and started offering new product offerings which were more or less repackaging of old products. A few startups also tried entering the market and made some decent impact initially but none of them were able to even grow to a fraction of the size of the big corporations controlling this market.

On D-day suddenly a huge chunk of their clientele disappeared. Generally, old people are the biggest clientele of this industry. Old people generally have more medical issues which is beyond their control, hence feel helpless and hence become customers of this industry. Old people also generally have more time, post-retirement, and hence use the products of this industry more. Old people are also the biggest brand ambassadors of this industry. They not only introduce the products and services of the industry to their friend but also hard sell the products and services to the younger generation, especially their grandkids. The youngest generation is generally exposed to this industry by the grandfather generation. It works amazingly for this industry. Their biggest customers keep getting their youngest customers into the fold, who will continue to remain their customers for decades and bring in more customers over the period of time. In a way, it also works like a pyramid scheme of marketing practised in a few other industries.

This industry is one of the oldest industries and most of these major corporates have survived and prospered for hundreds and thousands of years. Suddenly after D-Day, they experienced their major existential crisis. They were not at the risk of going extinct, they are too big and too entrenched for that, but they faced the biggest erosion in their customer base. As their marketing plan was based on a pyramid scheme, the sudden loss of customers also meant a sudden loss of a big part of the distribution network. The loss of old people, the grandparents, meant that there was nobody to introduce and hard sell the products and services of these corporations to the new generation. If the new generation is not introduced to their products and services and if the old generation gets kicked off from the world early, the business of these corporations is crushed from all sides.

The salesmen on the corporate rolls, the branch managers, and most other employees of these corporations were generally senior citizens, especially the top management. Due to the sudden increase in business after the announcement and believing their own mis-selling, they did not prepare for the upcoming Tsunami. D-day switch-off of senior citizens hit these corporations like a tsunami as most of their top management, branch managers, and salesmen, suddenly disappeared from the face of the earth. The corporates did not get any time for knowledge sharing and handover.

Another big issue was that the remaining client base started questioning the effectiveness of their products and services. When the big announcement was made most of these big corporates tried to encash the sudden increase in demand. Many mis-sold their products and services claiming they can stop the D-day or neutralize/contain/mitigate the impact of D-day. D-day came and the impact was very brutal. Most of these corporates were not able to do anything. This led to disillusionment and disenchantment about the products and services in the minds of the clientele.

Another threat was the rise of a new startup, a new religion that started seeing Leo as their God. They started believing that Leo was the person who controlled their fate and their death. They started treating him like God. They started praying to him, they started loving him, they started fearing him. Leo loved to play God and loved the sudden massive increase in his devotees. A few smart people tried to become Godmen of this new startup and open their own shops selling products under the brand name of this new God. However, Leo did not support the opening of any independent branches and selling products in his name. He used AI tools and his social media channel Y to trace and weed out all such counterfeit shops that were selling products and services in his name and crushed them hard. He believed in direct selling without any middlemen, to maximize profits.

The Medical Industry

Another industry that dealt with life and death was hit by the tsunami due to the switch-off. The medical industry which included doctors, pharma companies, hospitals, pharma retailers, medical centres, medical test service providers, and nursing service providers was hit hard by the tsunami of D-day.

Unlike the God industry, there was better awareness of impending tsunamis in this industry. This industry is much better structured, has many listed companies, and is hence well covered by analysts and research reports. The corporates and businessmen in this industry are considerably better educated and they were aware of the writing on the wall and were preparing for it. However, you can never prepare enough for once in a lifetime (lifetime of the earth, not humans) event. It is impossible to predict with confidence how things will pan out.

The hit was brutal. Suddenly major part of the clientele disappeared. It used to be a joke among the businessmen of this industry, that humans earn all their life, save all their life for them. The major chunk of the earning and savings of the clients used to be sucked up by this industry before their death. The industry used to keep the client alive so that it can suck the last dollar from them. The client also knew that they couldn't take the wealth to heaven. So, they used to keep spending till the end of their wealth and the end of the next generation's wealth, to stay alive.

Humans toil all their lives to earn money. In the process embracing many diseases as a compromise to earn more.

However, post-retirement around 65, they used to realize that they had made the wrong compromises. They realize that money has limited utility and health is a bigger wealth. Then they start spending on their health till the end of their wealth. Humans typically end up working for two industries all their life. One is the real estate industry which sucks a major part of their income and then the medical industry which keeps sucking till the end of their life or till the end of their wealth, whichever is earlier. Old people who are lucky to die earlier can leave some wealth to their next generation, people who live longer end up giving their next generation liabilities rather than assets in inheritance.

Leo and the other creators of the project Auto-off @ 65 blame this industry for pushing them to conceptualize the project. The creators credit the industry for the good work done for hundreds of years and for extending the life of humans for so many years. But they feel that the industry went too far. Now the life of humans is extended to such an extent that it is now becoming a threat to mankind. There is an irony in the argument but now in 2030, people can appreciate the irony better. In 2025, at the time of the announcement, people were not able to appreciate the irony.

Players of this industry including the doctors and corporates were one of the biggest critics of the project. The doctors are very well respected for their knowledge and when these doctors were criticizing the project, we were agreeing to them. However, when they were criticizing the project, they also had their own interest in mind they knew this project would hit their industry very hard.

On D-day suddenly a big chunk of their customers disappeared from the face of the earth. Although only 10% of the human population disappeared from the Earth, for the medical industry the sudden fall in business was more than 50%. The impact was brutal and most of these companies went into red due to lower capacity utilization. The industry also lost many doctors, as doctors generally do not retire and keep working till the end of their lives. The loss of doctors was bad for the industry but also helped in bridging some gap between demand and supply. Due to the switching off of old people, a huge oversupply was created in the market. Switching off of a big percentage of doctors helped in reducing the gap.

Some companies were hit harder than others. Some companies that targeted old age diseases, like Alzheimer's, cancer, brain stroke, and heart stroke were impacted more. Others like pregnancy-related services, childbirth-related services, neo-natal, etc actually gained from the switch-off.

There was a steep fall in funding for the sector. The funding for research that used to come from the government and other private trusts and organizations started to dry up. Now people were not worried about cures for many diseases as anyway, everybody had to die at 65. Elongating life with new medical discoveries was no longer an option.

The biggest gain from the fall in medical expenses was for the state exchequer. In most countries, the government subsidized healthcare for the poor. Now government is able to substantially save on these subsidies. Countries like Denmark, Sweden, France, and Germany which have high ratios of greying populations and provide free medical expenses were the major gainers. The governments saved a lot of money which they can now invest in more productive uses.

Call from Mars

Laxman was delighted when Leo announced that a call was possible between Earth and Mars. He was desperate to talk to his mother, Sumitra, and confirm she was doing fine. Earlier, after hearing rumours, he was sceptical about whether Mother was still alive or not. He was sceptical about whether Leo had actually colonized Mars or if they were just fooling people. But now that Leo has announced a video call between the two planets, he is pretty confident that Leo has not been fooling around. His mother is actually on Mars.

It took Laxman more than a week to figure out exactly how he could talk to his mother. A lot of people were desperate to talk to their relatives on Mars and it took SpaceY some time to streamline everything. Laxman was elated when he heard his mother. He was more elated when he saw her. 30 minutes of video call was free and he thought that was more than enough. His calls with his mother, when they were in Mumbai and Bokaro respectively used to barely last 10 minutes. Eyes were moist on both ends when they started talking, and everybody in the family joined on the video call to greet Sumitra. She was equally elated to see all of them again, she had lost all hope. She was specifically ecstatic to see her grandchildren again. 10 minutes passed in a flash. Laxman was conscious that his other siblings would also like to talk to their mother and that he should not use up all the 30 minutes. They decided to disconnect and connect again next week. Laxman was very curious about life on Mars, he had so many questions in mind, and 10 minutes would not be enough for that. But he is glad he managed to see his mother;

she appeared fit and fine. His queries about Mars can wait for next week. Sumitra needed to get a chance to talk to her other offspring too.

Now that two-way travel to Mars has opened up for people of all ages, he would love to visit his mother there. But $100K is a big amount, he does not have that kind of money. Added to that, travel to Mars still takes months. Two-way travel will require more than 6 months. He cannot be away from Earth for so long. He has a career and family to look after. He tells himself – maybe sometime in the future, it's been less than a year since mother left, even earlier they used to meet only once a year during Diwali holidays. He was very glad that he would be able to talk to her and see her, every week. He was very grateful to Leo for the same.

Jane Cash on the other hand was not happy to receive calls from her father every now and then. Mickey being a co-conspirator along with Leo, had no such limitation on call frequency or duration. He had nothing much to do on Mars and called up Jane frequently to advise her. He wanted to still run the show, now not from behind the screen, but from another planet. It is very difficult for powerful men to give up power. Mickey managed to give up the planet but is unable to give up his lust for power.

Jane is feeling a monkey on her back and she wants to get rid of it. She has stopped taking his calls. She makes her secretary politely convey that she is busy with some Presidential work or other. The world has completely changed since when Mickey was here. Most of Mickey's contemporaries have also been kicked out of the Earth. Mr. Turin and Mr. Lian, the two biggest problems for any USA President are also kicked out. Mickey used to guide Jane on how to handle them, but now that they are not there, Jane needs no guidance. Their successors are totally different animals, about whom Mickey knows nothing, hence he could not be of any help. Jane has now spent more than 18 months in the post of the President of the USA. She is now pretty confident. All the oldies in the capitol have also been kicked out. So, there is nobody who is throwing his weight around. She is no longer a lightweight. She is no more a newcomer. Compared to others, she is a veteran now.

Next week, Laxman again spoke to her mother. He learned from her about the facilities there, and her daily schedule. He enquired about her health and the food. He was glad that vegetarian food options were there, he knew his mother would die but not touch non-vegetarian food. He was pleasantly surprised that most of the food was vegetarian as there were no animals on Mars yet. Agriculture was the only source of food. He was glad to know that his mother had no complaints and she had adjusted well to such a different environment. A different planet. Laxman knew his mother was a strong lady with limited needs. She can adjust anywhere if she puts in an effort. He was very happy she is taking it positively and has adjusted well.

Chapter 7:
2035: The New World Order

Democracy

The world changed in a blink on 1st January 2030. Since then, things are evolving at a rapid pace. The world is evolving at a rapid pace. Somebody who left the Earth on 31st December 2029 will not recognize the Earth if he lands back today.

The biggest change was the change in the resistance to change. And that changed everything. Old people used to resist change. As you grow older resistance to change increases. It increases because your ability to learn and change decreases. Older people were in power all over the world and hence the world was resisting change. With the kicking out of old people from the planet Earth, younger people got control of all decision-making positions. They were much more amenable to change. Hence the pace of change in the world increased.

Laxman who is a tech entrepreneur has been trying to convince the bureaucrats and politicians of his country to move to an online voting technology. India spends more than five billion USD on each national election and an equivalent amount on state and local elections. More than 10 million people are required to manage these elections and still voter turnout in elections is less than 70%. Online voting technology is superior to offline voting systems in many ways. It not only saves cost and time but also increases voter turnout substantially. It is also much more secure and transparent than offline voting systems. Laxman has built a very secure and high-tech online voting platform, tested, approved and certified by the Government of India, way back in 2015. Till 2030 he faced a lot of resistance in the adoption of the technology especially for government elections.

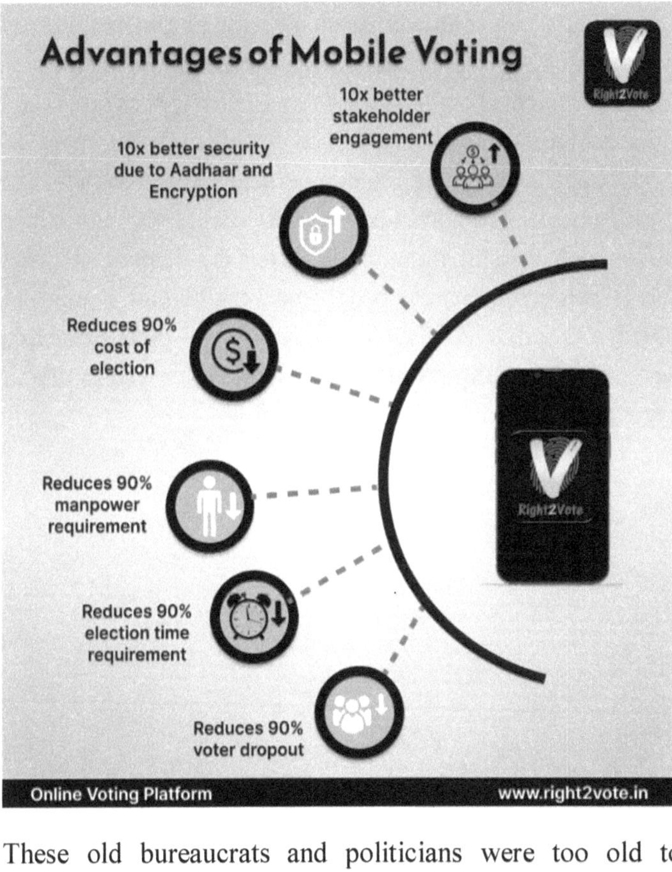

These old bureaucrats and politicians were too old to understand technology and hence were fearful of embracing it. Things suddenly changed after 2030. Once the old people got kicked out of the planet, younger people took charge. They were much more amenable to technology and change. This led to the quick adoption of online voting technology in government and other elections. From small to big, all types of organizations have completely embraced online voting across the world. Nobody uses paper voting anymore, in fact, now the world uses paper mostly for toilet paper only.

The political environment has changed at a rapid pace across the world. People are much more aware and conscious of their own rights. They now insist on their **Right2Vote**. Once old autocratic leaders were kicked out of the Earth, many of these countries converted to democracies. Some were less in spirit, some more in practice but most started involving their citizen in decision making. Even Russia is a democracy now! Direct democracy has also become very common with the advent of online voting technology. Many countries have direct citizen voting on major decisions. Even TedLand and LeoWorld are run like a democracy now!

Energy, Transportation, Mining, Global Warming

Mankind has almost given up fossil fuels and moved to renewable energy. Leo's speech in 2030, which threatened to kick out more people from the face of the Earth, helped. Some people were able to appreciate the threat of global warming, some people were able to appreciate the threat of Leo better. Either way, people adopted renewable energy in a big way. Mining of lithium, rare earth, and other minerals on Mars also helped. Battery technology became much more robust and efficient and hence eliminated the time-of-day problem that was holding back solar and wind energy. Overall, the Earth has adopted renewable energy at a very fast pace, which has substantially reduced pollution on the Earth. All automobiles in the world now run on electric energy. Leo's electric car company, Nikla, continues to be the leader in transportation on Earth. They now also operate airlines that provide on-Earth and beyond-Earth transport services. These aircraft are now somewhere between aeroplanes and spacecraft. Many of these aircraft fly to space and back to cut down travel time between two cities in two corners of the world. Leo is not alone; many other companies provide similar services. Similarly, the world has extensively invested in a network of hyperloops. They hyperloops crisscross continents and have made the Earth very small.

Beyond Earth also, the transportation infrastructure has made rapid progress. All spacecraft, without exception, are now using electric energy and battery technology. Travel to space especially Moon and Mars have become very common. Not only SpaceY and Green Origin, but many national and private companies now offer space travel services. India's ISRO has direct flights to Mars every day of the week from different Indian cities. Many other Indian startups are also planning flights to the Moon and Mars. The space travel time and cost, both has come down substantially. It has now become very common for middle-class families to travel to space every two to three years.

Mining & manufacturing activities on Earth have also reduced substantially and such activities have now been shifted to other planets. This has also helped in controlling global warming. People have become much more conscious of global warming and have started practising conscious consumption instead of conspicuous consumption. Apart from Leo, many other leaders of the world are now very pro-climate and government policies are very focused on environment management.

Education and Work

One of the major changes that has happened due to Project Auto-off @ 65, which was not that obvious way back in 2025, is in work and education. As old people have left the Earth, younger people are getting more opportunities. Now people are joining the workforce much younger.

Earlier in many countries, there was the problem of child labour. That has been completely eliminated. Most of the manual work which is straightforward and does not involve complexities has been offloaded to robots. 12 years of education has become mandatory and free almost everywhere in the world. Most people join the workforce immediately after 12 years of education at the age of around 18. These 12 years of education are quite extensive and cover much more than they would cover in schools earlier. However, it is much less taxing as methods of teaching and tools have substantially improved making it easier for students to understand and learn faster. Students now decide and study what they want to study at the secondary school level.

Only students who are very academic and research-oriented study beyond twelve years of schooling. Even at the age of 18 people get good corporate jobs. Even researchers, doctors, and academicians complete their studies by 21 years of age and start earning. The world realized that unconsciously more years of academics were being added across the world just to delay people from joining the workforce. As job opportunities were less, countries were forcing their youth to spend more years in college. That was not adding a lot of value to the students but wasting their most productive years. Earlier, students used to study only for a piece of paper, called a degree, which used to help them get jobs.

People started becoming independent at an earlier age of around 18 to 21. This gave them a working life of around 40 years, which is much more than earlier. Many people have started taking early retirement at around 55 years so that they can enjoy ten years of retirement before getting switched off.

Job opportunities are much better. Work is generally much more meaningful and most of the menial work has been delegated to robots. A lot of job opportunities have been created by the space, travel, and tourism industry. A whole new world of opportunities has opened up. There are many companies looking to colonize different planets even beyond the solar system.

Most of the countries have also introduced Universal Basic Income (UBI) for people who do not have any skill sets or who do not want to work. Very basics of life like food, clothing, shelter, medical, and education are manageable with UBI. Anybody who wants more from life needs to work and earn a good life. UBI has helped ensure a decent quality of life for everybody. It has also reduced the pressure on the job market.

Married Life and Family

There was a problem of child marriage in many countries and many communities. That has substantially come down with mandatory and free education. That has also come down substantially due to Universal Basic Income. Most countries have come up with laws restricting marriage below 21 years of age.

People have started marrying around 25 years of age and having kids around 30 years of age. Five to seven years of working before marriage helped people become independent and make their own independent decisions at the time of marriage. The influence of parents in decisions relating to marriage has substantially come down with increased independence.

As people are having kids between the ages of 25 and 35, kids are now healthier. Parents are also in a better position to provide for their kids. The gene deterioration problem, one of the biggest reasons behind Project Auto-off @ 65, has been arrested.

As people get married at around 25 and have kids around 30, their kids become independent when their parents are around 48 and get married when their parents are around 55. Age between 55 and 60 has become the perfect time for people to retire and enjoy their retired life.

As kids are becoming independent earlier and the maximum age of self is restricted to 65, there is not too much pressure on people to save for retirement. Universal Basic Income is also helping people to be less worried about finances and savings.

Parents are getting switched off when individuals are around the age of 35. Most parents are working till 55-60 and have only 5-10 years of retirement which they have to plan for. Hence there is no responsibility or pressure on people to take care of their senior generation.

Certainty of the timing of death is helping people live a much more planned and structured life. It is also helping people live a stress-free and anxiety-free life.

Organizations & Management

Organizations have become much younger due to the kicking off of all the sixty-five plus from the face of the Earth. From the biggest corporations to the smallest companies, associations, societies, trusts, unions, and organizations are becoming younger. Management is becoming younger.

Senior people are abdicating their thrones to younger people sooner. People nearing 65 anyway have no option to cling to their posts, but people in the age group of 55 to 60 are also abdicating earlier. There is no false notion they can hold on to the posts forever. They now want to retire early and live their life. Living life has become much more important than clinging to power or clinging to life.

Certainty of the date of death has changed humans forever. Unlike earlier, the lust for power and control has reduced and people now are more focused on enjoying life and if possible, retiring a few years early to enjoy more.

Parents are giving up control of their family business to their younger ones sooner. Businessmen also have a retirement age now. Politicians have no option but to retire around 60. Professionals like lawyers, doctors, chartered accountants, actors, directors, writers, commentators, singers, and cooks who never used to retire now have no other option. This is creating space for the newer generation and the new generation is getting opportunities earlier.

The new generation has breathed new life into these professions. The new generation has new and fresh ideas, is willing to adopt technology faster, and is willing to take more risks. Earlier all the decision-making and management in these professions were monopolized by the senior generation. Now the new generation is calling the shots and doing a much better job. This is making these organizations perform much better than in the past.

Law and Order

The law-and-order situation has substantially improved in the world. Technology has substantially helped in the fast delivery of justice. **Time2Justice** used to be the biggest issue with judiciary across the world. It used to take years for human judges and human courts to deliver justice. The judicial system used to suffer from the problem of limited number of judges, limited infrastructure, slow and long process of justice. The advent of AI and Robo judges has made the process of justice delivery much fairer, more transparent, and quick. Speedy delivery of justice has again instilled fear of the law among criminals.

SpaceY, Green Origin, and other space tech companies have put millions of satellites in space. Satellites have become very powerful and omnipresent. They cover each and every corner of the Earth. They are recording videos and images of every activity of humans across the world from each angle, they are like multiple CCTV cameras in the sky. AI is being used to automatically process images and videos to trace law and order issues. If any law-and-order incident is flagged by the AI systems, the system itself initiates the case and delivers justice. It is something like an automated road ticket practised in a few countries in 2025.

Some care has been taken to ensure the privacy of people, but most people do not care about it much. Anyway, these images are being processed and flagged by AI systems and not human beings. People are fine with paying a small price in the form of privacy for this automated AI-based judicial system.

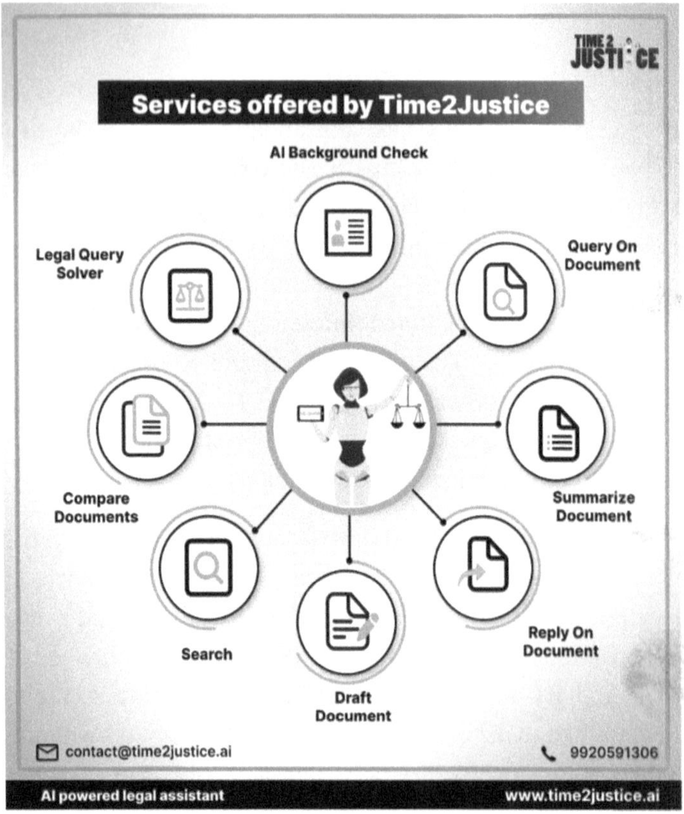

Also, due to universal basic income, there is a fall in law-and-order incidences.

Religion & Customs

The Tsunami that hit the God Industry made the business around these religions and the Godmen weaker. The certainty of the timing of death eliminated the fear of death. Elimination, or reduction in fear of death made the influence of religion weaker on the society.

People still practice their religion but now religion is a private affair. Most of the countries in the world are now secular and the constitution of the respective countries is above any religious book. There are still a few exceptions but by and large religious fundamentalism is not as big a problem in 2035 as it was in 2025.

People generally seek solace in God and religion more when they feel helpless. When people are young, healthy, and in control, religion is not that important to them. But when they grow old, sick, diseased, and helpless, God and religion are sought for support. Project Auto-off @ 65 has cut off 20-30 years of the most helpless phase in human life. Elimination of the helpless phase has also reduced the importance of religion in the life of humans.

Religions are not eliminated. They are still there and contributing positively to mankind. Now, people are not trying to impose their religion or faith on others. Satellite and AI-based law and order systems are also helping against forceful conversions and other form of religious imposition.

Old customs and traditions that were continued over a period of time without followers knowing the reason behind them are dying. The tsunami that hit the God Industry in 2030 suddenly eliminated the old priests and old people who were repositories of these customs and traditions. The Tsunami did not even give enough time for proper knowledge transfer or handover.

Mandatory and free education and upliftment in quality of education due to technology also helped in restricting Godmen and God industry from having too much influence on society.

When humans started travelling and colonizing planets in outer space, people were able to look through the hollowness of some religious teachings. In one specific case, a Priest, to scare his customer, told him that *Shani* (Saturn) was troubling his career and creating health issues. The daughter of the customer, who overheard the conversation, reacted – *"Papa, let's travel to Saturn for this holiday and get the blockage removed."* People started realizing that humans have now advanced much beyond what these old religious books ever imagined. They realised that it does not make sense to keep following some hundreds and thousands of years old religious books when the world has changed so much.

Tourism

Tourism is now one of the largest industries in the world and also the fastest growing. With the elimination of fossil fuel and the advent of renewable energy along with robust battery technologies, the cost of energy and hence transportation has dipped substantially. The advent of driverless cars and aircraft has made travel very comfortable. Advancements in space technologies have made travelling within Earth also much faster and cheaper.

An increase in disposable income in the hands of people due to savings in medical bills has helped people increase discretionary spending on tourism. Certainty of the end date of life discourages people from postponing their pleasure trips. The reduced number of people on Earth has increased the overall prosperity of the world. Faster transportation has made the world a smaller place.

Pleasure trips to outer space are now common. TedLand is now a sought-after destination for weddings. LeoWorld has become the hub for space travel. SpaceY is still the leader in space travel and they have made Mars their base. They are connecting newer and newer planets via their base at Mars.

As people are retiring between 55 and 60, people are using the last 5 to 10 years of their life in travel. People are both healthy and wealthy when they retire, plus they do not have any responsibilities to their parents or children. This phenomenon is encouraging people to take more pleasure trips after retirement.

Tourism is also contributing a lot to employment generation in the world. The hospitality and travel industry employs a lot of people. As humans explore more and more places in outer space, the demand for people will only increase.

Medical

The medical industry has collapsed compared to 2025 but is doing somewhat okay after the shock of 2030. People have become health conscious and diseases in the world have reduced. Still, the medical industry is finding a lot of work in preventive care.

People below the age of 65 also need medical services and the medical industry is now focused on such clientele. There is an increased requirement for medical studies for the environment outside of Earth. Most of the medical research, medicines, and equipment built till 2025 were created keeping in mind the Earth's environment. Now that humans are exploring new planets; the medical industry has to keep pace with the developments and find solutions for different environments.

Medical facilities are now almost free in most countries. In some countries, it is funded from tax dollars, in other countries they have made standardized medical insurance plans mandatory for all citizens. As the premiums of these medical plans have substantially come down, due to reduced risk of diseases, the cost of medical facilities has eased for citizens.

Government Finances

Government finances across the world hit two major bonanzas. First was the savings in medical expenses. Most governments across the world used to spend a substantial amount in providing their citizens with either free medical services or subsidized medical services. With the kicking out of senior citizens from the face of the earth, these expenses were reduced substantially.

The second major bonanza was in the form of reduced retirals payout. Most governments across the world have not anticipated and not planned for the steady increase in average life expectancy. Over the years governments have promised their government employees, their armed forces personnel, and others a certain rate of retiral. These promises were based on a forecast of lower life expectancy. However, with modern medicine, this increased much more than expected. This led to government finances going in the red.

Rising medical bills, rising retirement benefits, and increasing interest costs had created a snowball effect that was forcing most government budgets into deeper red. Even the developed countries and the performing countries like the USA and developed nations in Europe were facing this issue in 2025. Increased government borrowing due to a deficit budget led to more borrowing by the government and hence rising interest rates across the world. This added to the snowball effect. It became a vicious cycle where each factor was aggravating the other factor leading to more and more deficit budget. Medical and associated expenses during COVID-19 only accelerated the snowball.

The bonanza of reduced medical bills and reduced retiral bills helped countries break the vicious cycle and stop the snowball. Government finances went back into black which reduced the need to raise government debt. When government debt was no longer crowding out private debt from the market, the private debt became cheaper, and hence private investments increased.

The economies across the world started booming due to increased private investments. It was also supported by a boom in the space exploration business, a fall in prices of important minerals due to import from outer space, efficiencies in business processes due to AI, new technological innovations, fall in energy costs due to renewable energy. The vicious cycle converted into a virtuous cycle. The snowball effect started acting in the positive direction.

Surplus budgets enabled many countries across the world to start Universal Basic Income (UBI). UBI with negative budgets would have aggravated the vicious cycle. However, UBI with a surplus budget intensified the virtuous cycle. UBI significantly helped the poor and marginalized. This started helping solve the problems of malnutrition, lack of education, and employability. It also saved people from getting caught in a vicious cycle of poverty due to sudden medical or other emergencies. Reduced number of dependents, and lower medical bills due to Project Auto-off @ 65 helped the poor people escape from the vicious cycle of poverty.

The impact of project Auto-off @ 65, UBI, and a booming economy was increased spending power in the hands of individuals. A booming economy created a virtuous cycle of increased disposable income in the hands of individuals, leading to increased consumption and investment.

Personal Finance

Project Auto-off @ 65 significantly helped the personal finances of individuals and the budget of families. Reduced number of dependents, lower medical bills, short retirement life, early start to productive life – all these factors substantially improved personal finance. Then the second level impact like UBI and the booming economy kicked in which accelerated the pace of improvement in personal finances.

Reduced number of dependents: This was one of the most important positives from the project. Earlier as experienced in the case of Mr. Lee from China, there were as many as 16 dependents on a single earning member of the family. He had no option but to get caught in a vicious cycle of poverty. It was impossible for him to climb out of the ditch of poverty with the weight of sixteen people on his shoulders. The project has switched off twelve of them. People like him now need to look after only himself, his wife, and the two kids. He also does not need to plan and save for a long retirement. His kid will also join the workforce at an earlier age. All this has helped him escape from the vicious cycle of poverty. There are many like him. Now they can spend much more freely, have a better life, and invest in their kids' future by giving them a better education.

Lower medical bills: In 2025 there was a popular inference of economists for the middle and lower middle class. The inference was – '*each of these middle-class families is one medical event away from poverty.*' With fourteen dependents in the age group of 65 to 90, not having a medical event was a 'black swan event' and not the other way around. These old dependents will continue to have medical issues and will drain the family budget. It was impossible for a family to spend on anything else. With the switching off of these people above the age of 65, the families' spending on medical expenses came down drastically. A fall in medical expenses increased the disposable income in the hands of the family which they could spend on discretionary consumption.

No debt trap: Sudden medical events used to force middle-class families into debt trap. They ended by borrowing at a high rate of interest during such emergency medical events and then rest of their lives they used to keep paying the interest and principal. Before they can pay off the first loan, they will have another medical emergency for which again they will have to borrow. For each borrowing, the interest rates kept rising as the creditworthiness of the person kept falling. Soon families find themselves in such a situation that they have to take a new loan to pay off the old loan. This is a debt trap and there is no way out of this trap. Due to emotional reasons, families cannot cut down on the medical expenditure of their parents and grandparents. They will keep borrowing till somebody is willing to lend them. No good person would let their parents die because he does not want to borrow. Auto switch-off of senior citizens saved the children from the guilt of letting their parents die and saved them from debt traps. Leo and the other conspirators did a great service to them by taking away their guilt. They took the blame and abuse but saved millions from guilt and debt traps.

Shorter retirement life: Earlier people had to plan for a long retirement life of thirty to forty years. Not only for themselves but also for their spouse. Now the retirement period is only five to ten years. Shortening of retirement period reduced the pressure on families to save more. People could live their lives without worrying about the future. This helped people to spend more. The increased consumption, charged up the economy leading to more and better jobs.

Early start to productive life: Due to the kicking out of ten percent of the population from the face of the earth, new opportunities opened up for the younger generation. As old people were not vacating positions, the younger generation used to have limited opportunities. Therefore, the younger generation was forced to spend more years in colleges to earn more degrees which provided minimal incremental benefits. It was a double whammy. People had to spend a lot for college and higher education and also their income used to start three to four years late. Many times, the younger generation had to borrow for the college fees which used to keep pulling them down for several years. Now with an early start to the career, savings in education expenses, and four to five years of savings before marriage, the younger generation was very well off.

No loans, no EMIs: In 2025 it was common for young men to take huge amounts of loans including education loans, car loans, housing loans, and personal loans. They used to start working around 25 years of age and used to immediately get married. With the additional responsibility of the family, they had to spend on car, house, marriage, travel, etc. All this was possible only with loans. All these loans meant many years of Equated Monthly Instalments (EMIs) and little savings.

Many families used to get debt-trapped. Before they could pay off the loans, their parents used to retire and become dependent on them. Mr. Lee's family and most other families across the world were trapped in this cycle. Project Auto-off @ 65 broke that trap and gave freedom to people.

Booming economy: Project Auto-off @ 65 triggered a virtuous cycle in the world economy. Lower government deficit – lower government borrowing - lower interest rates – higher profits and investments in the private sector – higher tax collections and government income - Increased government spending – increased investments – more jobs opportunities and higher income – additional tax collection – UBI – higher disposable income – more consumption and saving – more taxes.

This virtuous cycle pulled hundreds of millions of people out of poverty across the world. Today nobody can deny that Project Auto-off @ 65 has been a huge success. Today each and every human will agree that the project has improved the quality of life of the humans on the Earth tremendously. Today everybody would agree that the three conspirators, conspired for a better future, and they were wrong in cursing them. The creators of the project played God but for the betterment of mankind and not otherwise.

Banking and Finance

The banking and finance sector also had a deep impact due to the project Auto-off @ 65. The project led to savings in government expenditure on medical and retirals which in turn led to a surplus budget for the government. The surplus budget led to a reduction in government borrowing which led to a reduction in interest rates in the economy.

Reduction in interest rates in the economy led to higher profitability for the corporate sector which meant more investments by them. Higher profitability for corporates meant better recovery of loans by the banking sector. Higher investments by the corporate sector meant more business for banks from the corporate sector.

Improvement in the private finances of people substantially reduced the requirement for personal loans and other loans taken by individuals. There was a steep fall in demand for student loans, car loans, housing loans, personal loans for marriage, travel, and medical emergencies. High-cost customer finance and buy now pay later industry also crashed. However, the fall in demand for individual loans was more than made up by loan demands from large and small corporates. Small and medium enterprises grew very fast in the last 5 years and have become a large contributor to the bank's balance sheet both on the asset side and the liability side.

The economy is booming, per capita income has substantially increased and so has household savings.

Corporations are doing well and the investment cycle is rising. All these have contributed to the growth of banks which have become much bigger and more profitable now. Even banks that were earlier struggling with their profitability and liquidity ratio are now healthy and prospering.

Communication & AI

The world has seen a mad rush to space in the last ten years. Millions of satellites are now dotting the space. Apart from the big players like SpaceY and Green Origin, many other companies and countries have launched hundreds of satellites in the last ten years. Communication had become cheap and easy in 2025 itself; it is a hundred times better today. Communication had become high quality and reliable in 2025 itself; it is multiple times better today. Anybody and everybody can afford satellite phones today which was very rare in 2025.

Today's world cannot be imagined without satellite communications. Half of the equipment on the Earth is remotely controlled via the internet or by AI. Self-driving cars and drones with capabilities to transport humans are common today. One of the biggest risks today for mankind is the dependence on these communication systems. However, several layers of redundancy have been built into the system and the communication networks are very robust. Satellite communication has reduced the dependence on wires and towers, making physical sabotage of communication lines difficult.

Artificial Intelligence (AI) has improved by leaps and bounds in the last decade. AI is now controlling almost everything in the world. AI has made work tremendously easy and efficient. AI has accelerated the pace of growth of mankind substantially. However, dependence on AI poses a new and evolving threat to mankind. AI is the best friend of humans today but it is also the biggest threat to mankind in the foreseeable future.

‘Live it up’ industry

An altogether new sector has evolved overnight. Psychologists have named it the ‘live it up’ industry. Humans now know that they will get switched off exactly at 65. Now there is no false notion that they can live till eternity. Earlier, humans had stopped living and were more focused on surviving. People would avoid taking risks so that they do not damage anything, and they can live longer.

Now they know that they cannot live longer than 65 years. Now they do not want to be switched off with a whimper. They want to go off with a bang. Humans are living up their lives, especially in the last five to ten years. Humans are making sure that they complete their bucket list before they are 65. They are no longer postponing because they know that death can’t be postponed.

Humans are taking up extreme sports like anything. Especially people in the age bracket 55 and 65. Earlier less than 1% of the human population used to participate in extreme sports. Now the percentage is higher than 50% in the age bracket 55 and 65.

Humans are doing bungee jumping, sky diving, deep sea diving, paragliding, base jumping, speed skiing, white water rafting, scuba diving, rock climbing, and mountain biking. Every day a new extreme sport is being invented. Humans are thinking *‘Anyways we have to die at 65, so why not test our luck and enjoy the thrill of these extreme sports? Anyway, what's the risk? Dying a few days earlier than 65th birthday?'*

The old age homes and Ashrams have been replaced with these 'live it up' centres. These centres offer day plans and even long-term plans. As long as 10 years. And there are people who are signing up for these long-term plans and booking their seats many years in advance. People enjoy learning skills like wave surfing, mountain climbing, scuba diving, and skiing at the sunset of their lives.

There have been a few cases of people doing negative activities in the last few days as they know that they would be switched off by the time law catches up with them. Few tried to settle their old score. However, such incidents are not as many as feared. This might be due to many reasons. People do not want to spoil their legacy and memory for their loved ones. People do not want to spend the last few days of their lives in negativity or jail. In many cases, individuals, when young and in a fit of anger, plan to murder their rivals when they reach closer to 65 but when they reach closer to 65 the rivalry stops meaning anything to them.

Satellite and AI-based law enforcement also helped mitigate this problem.

Chapter 8:
The Space Conquest

UN – Rights over Mars and other planets

The space sector has made tremendous progress in the last ten years. The sector is not monopolized by the likes of SpaceY and Green Origin alone. There are many other startups that have now become very big. Apart from NASA of the USA and ISRO of India, many other countries have well-developed space programs and are launching new projects regularly.

These companies and countries have started fighting over control of land on other planets. The fight has intensified so much that a requirement was felt for getting all countries to meet and discuss a solution. The matter was taken to the United Nations (UN) for debate, discussion, and final resolution.

Leo and Ted being two of the biggest players got the opportunity to speak their point of view first. Ted has left the Earth and is now a resident of the Moon and Leo is on a visit to Mars. Hence, they joined the meeting via video conference. As expected, Leo was demanding exclusive rights over Mars as SpaceY was the one who colonized Mars first. Leo also reminded the UN that they have agreed to give SpaceY exclusive mining rights over Mars. As no end date was specified in the permission, Leo insisted that it means they have exclusive rights till eternity.

Ted's position was no different. He wanted exclusive rights over the moon. He not only wanted exclusive mining rights but also exclusive rights to land spacecraft on the moon. However, he also wanted permission to land Green Origin's spacecraft on Mars and also wanted mining rights on Mars.

Most other countries wanted access to both the Moon and Mars. They were against any exclusive rights.

Leo, knowing he had a lead over all other players in Space exploration gave a formula that whoever reaches a planet or celestial body first should get exclusive rights over that planet or celestial body. Everybody else, including Ted, fought against Leo's proposal but Leo continued to argue in favour of his proposal.

After a major heated debate, when Leo was not conceding any ground, Mr. Raju, the representative from ISRO of India, who was very quiet until now, stood up and spoke. He said – *"As Mr. Leo is not ready to relent from his suggestion, let's agree that the first organization on the planet will own the planet."*

Everybody else in the room went quiet. Leo was ecstatic. Leo was conscious that ISRO is highly respected in the space community and with ISRO concurring with him, it would be difficult for others to oppose.

Mr. Raju continued, *"ISRO's rover reached the Mars environment much before any rover or spacecraft of SpaceY reached Mars. We have shared rims of data from our Mars mission with mankind which has helped SpaceY and others to progress in their space mission. NASA and others have also reached Mars before SpaceY and we are agreeable to sharing the planet Mars with NASA and others who reached there before SpaceY. We would request SpaceY to hand over Mars to us with immediate effect."*

It was a game, set, match. People in the room started laughing. Leo went red. Even Ted started laughing. He loved it. Nobody has given it back to Leo like this in the last ten years. Leo argued that ISRO, NASA, and others have sent rovers to Mars and have not colonized it or sent humans to Mars. But all his arguments now appeared weak and without conviction.

Finally, he relented and agreed to a compromise. Based on further discussion, debates, and suggestions from ISRO, all parties agreed to a common formula.

It was agreed that there would not be any restriction on any company or any country to land on any celestial body. The organization that lands on these bodies would have the right to occupy and use the land for mining and other activities. They would also be allowed to build colonies.

Any other organization landing on the celestial body subsequently, would be allowed to occupy and use land which is not already in use by the organizations that have arrived earlier. Mere fencing of land would not be considered as occupied land. There should be real activities on that part of the land.

UN also laid down rules for ensuring that any original inhabitants of the celestial bodies would have first and superior rights over the planet. The rights of the original habitant should not be tampered with. UN should be informed immediately if any celestial body is found with any kind of living organism. Rules were also set to ensure proper use of the planet and avoid destruction and pollution on these celestial bodies. Rules were also laid for renting and leasing of land on these bodies between different organizations. Both Leo and Ted reluctantly agreed to the decision.

In the end, the UN also stipulated that these celestial bodies, where humans have made colonies, should operate like democracies, and the companies running the operations there should not operate like autocrats. Both Leo and Ted were taken aback. Being US citizens, they have been big proponents of democracy all over the World. They have argued against and mocked Russia and China for not giving their citizens the right to vote and right to decide their own fate. However, in space, both these people have been operating like Kings. Leo was the emperor of LeoWorld and Ted was king of TedLand. Now the tables have turned on them.

Leo and Ted contested it and tried to explain how they had spent billions to colonize these celestial bodies. They argued that they should be allowed to recover their investments and maintain total control over these bodies. UN reminded them they had already agreed that other space companies could land and build colonies in these celestial bodies. Hence, there is no way they can be allowed to dictate terms on the celestial body. Leo and Ted reluctantly agreed to implement democracy on Mars and Moon.

Ted and TedLand in 2035

TedLand has developed quite a bit in the last 5 years. Still, it has a very niche business model. Only around 1 million people are inhabitants of this small satellite. For these 1 million people the land mass of the Moon is too big. TedLand has not occupied even 5% of the total land mass.

TedLand has also opened up for tourism. They have built many amusement parks in TedLand which can be used both by the residents and by visitors. There are certain areas of TedLand that are exclusive for residents only.

The charges for entry and stay at TedLand have been steadily falling along with the cost of space missions. With the UN's recent resolution on the sharing of celestial bodies allowing other organizations also to set up colonies on the Moon, rates are expected to fall steeply.

Ted has built many premium facilities in TedLand over the years and sees no immediate threat from other colonies on the Moon but he has to be prepared for competition. After the UN meeting, he has taken a call to expand at a rapid pace so that he can occupy the maximum possible footprint on the Moon.

On the personal front, Ted is already 71 years old. His second wife will be turning 65 later this year and will be joining him at TedLand permanently soon. She has been visiting him regularly for the last 5 years and has been staying here for longish periods of time. She is the conduit between Ted and the Earth. Although Ted can communicate via video conference on Earth, and regularly does that, he can no longer visit the Earth. His wife's frequent visits help him stay better connected with Earth.

He has handed over control of his empire on Earth to the next generation. He takes out some money regularly from his various businesses on Earth and invests it in TedLand. After the UN resolution, he intends to expedite the expansion process.

Mr. Turin of Russia and Mr. Lian of China, the two most famous residents of TedLand have expired. After their respective proteges on the Earth stopped listening to them, they lost their influence on the politics of their respective countries. They grew very bitter and irritated. They used to confide with each other about their bitterness. Lack of power and status killed their lust for life. They started feeling their life at TedLand was meaningless. They longed to die. Soon God granted them their wish.

Leo and LeoWorld in 2035

Leo is now 64. He will have to move out of the Earth next year. He is still very fit and very ambitious. He is still considered to be the biggest genius on the Earth. He is managing an amazing number of businesses. What he has been able to achieve looks unreal and was not considered humanly possible before he achieved it. Many people believe he is not human. Some believe he is God; others believe he is from some other planet, an alien. A section of people on Earth believes that as he is an alien, he will not get switched off when he turns 65.

Leo has no such misconceptions. Although he might like to believe that he is God, he is well aware that he will also get switched off next year, if he does not leave the Earth. Hence is planning to migrate to Mars.

On Mars now he has created many colonies. Many of these colonies have better features than others. The earlier residents who did not pay to stay on Mars still have the basic colonies. People there are mostly content with what they have. Few of the newer colonies where residents have come in after paying the charges have more facilities and features. Leo is building colonies at each price point. Few of his colonies are competing with TedLand in terms of amenities. Leo wants to ensure that the most premium destination in space should be LeoWorld and not TedLand. Although Ted had a lead in terms of premium tourism, over the last couple of years Leo has managed to establish a few of his colonies as the most premium destination in space.

However, the colonies he is most excited about are the colonies where experiments on the evolution of mankind are happening. People in these colonies had lived on the Earth earlier and hence carried with them the knowledge that was available on the Earth. Leo had no way to wipe that out. However, he has made the environment very basic, without any amenities. He has created a setup of a jungle with plants and a waterfall so that the Adam and Eve of Mars could survive, but beyond that, he has not provided anything. They have not been provided even clothes, sanitation facilities, or basic tools. They have to make these on their own with the resources available in the jungle. The subjects of the experiment there know about fire and wheel from their memory from the Earth but now they need to create it themselves here. Leo has cameras all around these colonies and has a team that monitors the progress. His team shares with him the important excerpts of the videos regularly.

Leo these days is obsessed with colonizing other planets. He wants to make Mars the base for humans to travel to outer space. He has already launched manned and unmanned missions to many planets.

For his businesses on Earth, he has started involving eleven of his offspring in various businesses. Leo is superhuman and has been able to manage all these businesses on his own. His offspring, even eleven of them together will find it difficult to manage all the businesses, and he knows that. At times, he regrets not having more kids and is still open to the idea of having more babies. He is planning to make a couple of babies on Mars once he relocates there. He wants to ensure that from conceiving to birth, the babies or their mothers should not be exposed to the Earth's environment. He is thinking of running this experiment and checking whether the human child born on Mars would be subject to the limitation of auto-off @ 65 or not.

Laxman and family travelling to Mars

Laxman is doing very well in life thanks to the economic boom the world is experiencing as a result of Project Auto-off @ 65. Earlier he was not in a position to travel alone to Mars to meet his mother. Now he is planning a trip for all four of them. The cost and time to travel to space have substantially come down. Urmila and the children are equally excited about the trip. This would be their first trip to space.

All these years, Laxman has been in constant touch with his mother, Sumitra. It's been 5 years. Initially, he used to never miss the weekly call. But over the years things have normalized as if they were staying in different cities. Laxman loves to talk to her but is now not very sad that his mother had to leave the Earth. Video-calling facility has significantly helped them not miss each other. But still, he wants to visit his mother once and also visit Mars to experience it himself. He wants to experience what it feels like staying on Mars.

Laxman specifically booked for stay in the colony where his mother is staying so that he can stay close to her during the trip. He has also booked for one day in one of the premium colonies along with her mother so that she can also experience it. However, her mother has decided to stay in her colony only and she does not want to travel. Laxman understands that she does not want to see the richer colonies and feel jealous. They are living a simple, peaceful life without any materialistic lust and she does not want to disturb that. He appreciates that. Although her mother must have seen these facilities over video call with other relatives and friends, it is her choice and he respects that.

Laxman, Urmila, and their two children landed on Mars and went straight to meet Sumitra. The children were very excited to meet their grandmother. Sumitra became very emotional on meeting them, especially the children who are grown-ups now. Everybody's eyes were numb. Kids hugged their grandmother; it broke the dam. She started to weep. It was a very emotional moment for the family.

Laxman noticed that his mother has grown very old, however, she is still healthy enough to be independent. Many of her friends in the colony were not that lucky. The colony was filled with old people, many of them very old and very sick. Few of them were bedridden and dependent on others. There were robots to help out such people but still, it was a very sad sight. Micky Cash was one of them. He is very old now and bedridden. He is in a lot of pain and his suffering is not ending. He prays to God every day to make his pain go away. He prays to God every day to switch him off.

Laxman, Urmila, and the children were there on Mars for 15 days and they had a great time. Laxman was very happy to finally after 5 years spend time with his mother. The children were equally happy to meet their grandmother after such a long time. Sumitra told them all about her experiences on Mars. She explained in detail, with a lot of enthusiasm, how the planet has evolved over time. Laxman has heard some of these stories over call but he was enjoying all these conversations. He was glad that he is able to have a face-to-face conversation with his mother again.

A day before the scheduled day of departure, Laxman noticed that his mother was very sad. She knew that it would be very difficult for Laxman and his family to travel to Mars in the near future and she might not live very long. She was dreading the moment when Laxman, Urmila, and her grandchildren would leave. That was making her very sad. Laxman, Urmila, and the children spent the full day with Sumitra, playing ludo and other games. Around evening suddenly, she started feeling uneasy. Robot doctors were called in but they could not do much. She was gone in around 2 hours. Those two hours were very painful for Sumitra. And seeing the pain on his mother's face, it was very painful for Laxman.

Laxman, Urmila, and the kids were very sad. It was as if the mother was just waiting to meet her son and once, he was about to leave, she lost lust for life. Laxman was sad but also happy that he visited her before her death and he was next to her at the time of her demise. Laxman was also sad to see her go through so much pain in those two hours. Laxman wondered if it would have been better if she got painfully switched off rather than going through so much pain. Or he could have taken her along with them to Earth where she would have switched off automatically on entering the Earth's environment. But Laxman admits that was never a practical solution. It would have been equivalent to mercy killing which he would never have been able to do. Laxman is still confused about his feelings. He is still confused about what he would have preferred - mother getting painlessly switched off on the Earth when she turned 65 or mother travelling all the way to Mars, staying alone here for five years and then dying after a lot of pain. For himself, Laxman is more confident now that he would like to get switched off at 65, rather than buying a ticket to Mars.

Chapter 9:
Year 2045 AD

Year 2045 AD

Today is 28th March 2045. It is around 10 PM. A party is going on at Laxman's place. A lot of people have gathered for the party. It looks like a very big event. It seems all Laxman's friends and relatives are here. Many of his childhood friends, and college friends are also here. Many of them Laxman has not met for a long time. Everybody appears very happy and in a celebratory mood. The occasion deserves celebration. Laxman is about to get switched off in a few hours. The event is to celebrate a life well lived. The event is to celebrate a happy departure from the world. The event is an occasion for everybody to show how much Laxman meant to them and in the last moments of his life, it is a way to show their respect. Laxman is a Hindu. In Hindu custom, after the death of a person there used to be 13 days of mourning. Not anymore. Now people celebrate before the switch off. Friends and family gather before the switch-off, celebrate for a few days, and leave after the last rites. There is no mourning, only celebrations.

It's been fifteen years since the D-day, the first switch-off day. The world is very different from what it used to be fifteen years ago. Fifteen years back people were scared of death, people were scared of getting switched off. Fifteen years back death and switch-off were one and the same thing. Nowadays, humans look at switch-off very differently. Nobody fears getting switched off now. Switch off is the final destination, it is the final goal in a person's life. People look at switch-off as the final destination where they want to reach healthy, happy, and with a lot of friends. Switch off is the moment when a person reflects upon what he did right and wrong in his life. On the switch-off date, people hope to have a lot of people celebrating with them. The presence of a large number of people is a sign that many people love and respect him. It is a sign that he has lived his life well.

Laxman chose not to relocate to Mars, the Moon, or any other celestial body. He is content with a life well lived and he has no more lust to live longer. For years he was aware of his end date and he has prepared well. All his finances are in order. Both his children have completed their studies, got married and now living an independent life. His son is working in an Indian MNC and doing well. He is happily married and has two kids. His daughter has inherited his business and running it very well. She has managed the business extremely well and with her energy and zeal, she is taking the business to newer heights. She is also happily married with two kids. Laxman retired at the age of 55 and he handed over the business to his daughter. For 10 years he enjoyed his retirement and ticked off each and every item on his bucket list. From trying out adventure sports like skydiving to writing a book, he did it all. He did all he wanted to do ever in his life. He travelled extensively along with Urmila in the last 10 years. Now he is kind of bored of his retirement life. He believes he has done everything he ever wanted to in his life. He no longer has any lust to live longer. He preferred to be healthy, happy, and independent at the time of death rather than diseased, sad, and bedridden.

Laxman's wife Urmila is also about to turn 65 in 4 months. She is also going with the option of getting switched off rather than migrating to some other planet in space. Urmila has no concerns that she would be left alone after her husband's death. Anyway, she is also getting switched off in 4 months. She has already planned her remaining 4 months which includes another big party at the end. Laxman's and Urmila's children were aware of the switch-off date of their parents for a long time. They are mentally prepared. They also want their parents to leave the world healthy and happy without any pain.

Laxman and Urmila are not alone. Most of the people are now happily getting switched off. Switch-off does not scare people. Switch-off relieves people from old age and painful death. Switch-off relieves people from being dependent on others. Switch-off has relieved people from the anxiety and stress of death.

The world is a much better place today compared to 2025. Poverty has almost been eliminated; pollution levels have been controlled and the environment is cleaner and healthier. Human genes are not deteriorating anymore and mankind is now a habitant of multiple planets. People are living a shorter life but a happier and much more productive life. People are able to plan their lives and exit the world without suffering, pain, and regrets. Young people are getting more opportunities and they are growing faster. The world is progressing at a much faster rate. The project Auto-off @ 65 has been successful in all its objectives.

Acknowledgements

First and foremost, I would like to thank my 11-year-old daughter, Anusha Gutgutia, who is my editor, advisor, and mentor. She is one of the most voracious readers I have known in my life. In her 11 years on earth, she has already read more than twice the number of books that I have read in my 45 years. She loves fiction and her love for fiction inspired me to try my hand at fiction.

In the past, I have tried writing books at least five times and each time I gave up as I was not satisfied with the output. This time Anusha read the output after each chapter and gave me her feedback and suggestions. She guided me on how I could make it more interesting. The perspective of this young reader helped me improve my writing style.

In the last few years, she made me spend a fortune on her books. I managed to get multiple time payback on my investment by appointing her as my editor.

I also thank my brother Aaditya Gutgutia, wife Neha Gutgutia, sister Payal Agarwal and friends Ashish Heliwal, Chirag Daru, Deep Sood and Vijay for proofreading the book and giving me valuable suggestions to improve it.

I thank my wife Neha and my elder daughter Avika for bearing with me while I was engrossed in writing this book. I thank my parents for everything. Whatever I have learnt is because of them. Whatever knowledge I am able to share via this book is because of them. Very few parents give their children the freedom to think differently and follow their hearts. I am blessed to have such parents.

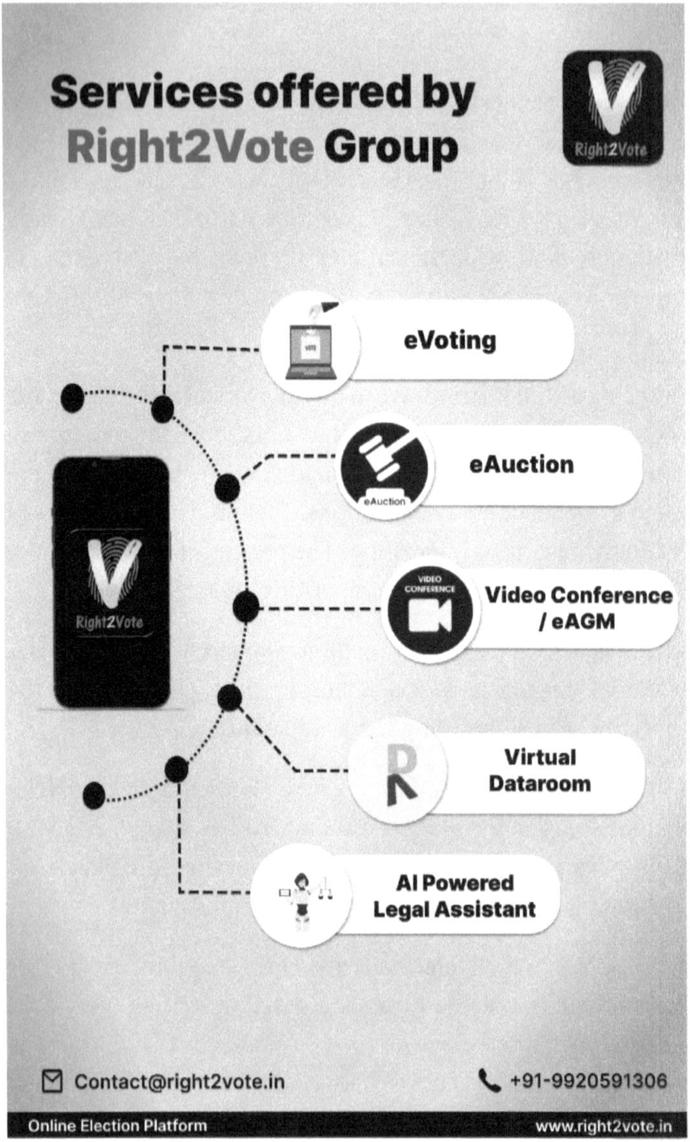
Services offered by
Right2Vote Group
Right2Vote
eVoting
eAuction
Video Conference
/ eAGM
Virtual
Dataroom
AI Powered
Legal Assistant
Contact@right2vote.in
+91-9920591306
Online Election Platform
www.right2vote.in

www.ingramcontent.com/pod-product-compliance
Lightning Source LLC
Chambersburg PA
CBHW050014170726
48282CB00025B/394
* 9 7 8 9 3 6 7 8 3 5 4 6 3 *